Cosmic Update

Multiversal Journeys

Founding Editor: Farzad Nekoogar, President
                        Multiversal Journeys

Editorial Advisory Board: David Finkelstein, Georgia Institute of Technology
                            Lawrence M. Krauss, Arizona State University
                            Mark Trodden, University of Pennsylvania

For further volumes:
http://www.springer.com/series/7919

Fred Adams • Thomas Buchert
Laura Mersini-Houghton

Farzad Nekoogar
Founding Editor

Cosmic Update

Dark Puzzles. Arrow of Time. Future History

Springer

Fred Adams
Department of Physics
Randall Laboratory
University of Michigan
Church Street 450
49109-1040 Ann Arbor
Michigan
USA
fca@umich.edu

Thomas Buchert
Université Lyon 1
Centre de Recherche Astrophysique de Lyon
CNRS UMR 5574
9 avenue Charles André
F-69230 Saint-Genis-Laval
France
buchert@obs.univ-lyon1.fr

Laura Mersini-Houghton
Department of Physics and Astronomy
University of North Carolina
Chapel Hill North Carolina
USA
mersini@physics.unc.edu

Founding Editor
Farzad Nekoogar
Multiversal Journeys
17328 Ventura Blvd.
Encino, CA 91316
USA
nekoogar@multiversaljourneys.org

The article by Lawrence M. Krauss and Robert J. Scherrer is reproduced by permission of its copyright owner.

Multiversal Journeys™ is a trademark of Farzad Nekoogar and Multiversal Journeys, a 501(c)(3) nonprofit organization

ISBN 978-1-4419-8293-3   e-ISBN 978-1-4419-8294-0
DOI 10.1007/978-1-4419-8294-0
Springer New York Dordrecht Heidelberg London

Library of Congress Control Number: 2011934680

© Springer Science+Business Media, LLC 2012
All rights reserved. This work may not be translated or copied in whole or in part without the written permission of the publisher (Springer Science+Business Media, LLC, 233 Spring Street, New York, NY 10013, USA), except for brief excerpts in connection with reviews or scholarly analysis. Use in connection with any form of information storage and retrieval, electronic adaptation, computer software, or by similar or dissimilar methodology now known or hereafter developed is forbidden.
The use in this publication of trade names, trademarks, service marks, and similar terms, even if they are not identified as such, is not to be taken as an expression of opinion as to whether or not they are subject to proprietary rights.

Printed on acid-free paper

Springer is part of Springer Science+Business Media (www.springer.com)

To My Parents
Farzad

Preface

> *Out for no reason this universe came to be*
> *Out of a place to be, this universe filled itself*
> Rumi
> 1207–1273

Many cultures throughout the history including the Greeks, ancient Egyptians, Indians, Chinese, and Persians fascinated by the night sky studied stars and the heavens for centuries. However, the study of the cosmos in the form of classical scientific astronomy using mathematical descriptions is traced back to early seventh century AD. From seventh to fourteenth century, Persian mathematicians and astronomers Kharazmi (780–850), Biruni (973–1048), Khayyam (1048–1131), Tusi (1201–1274), and Kashani (1380–1429) each contributed a lot to the field of astronomy. Their contributions were further developed and improved by European astronomers Copernicus (1473–1543), Galileo (1564–1642), and Kepler (1571–1630) over the next three centuries. Sir Isaac Newton's theory of gravitation revolutionized astronomical calculations by late sixteenth century (in 1687, Newton published his Principia). Newtonian mechanics made it possible to formulate the motion of all celestial bodies in the solar system and beyond.

New discoveries and theories within the last century have drastically changed our understanding of the cosmos. With the advent of Einstein's General Theory of Relativity and the observational discovery on the expansion of the Universe by Slipher, as well as Hubble's discovery of Hubble's law (indicating that far galaxies are receding from us) as early as 1920s, cosmology became a much more distinct science than astronomy. In 1922, Alexander Friedmann's solutions to Einstein's equations formulated the evolution of a relativistic expanding or contracting dynamic Universe. The more advanced models are now known as Friedmann–Lemaitre–Robertson–Walker (FLRW) models of cosmology due to many enhancements and contributions from other cosmologists.

From 1930s onward, the Big Bang theory formed the basis for explaining the expansion of the Universe. However, the original Big Bang theory endured three problems, namely, the smoothness problem, the horizon problem, and the flatness problem. The first problem asks why the matter is uniformly distributed in the Universe. The second problem concerns the large-scale uniformity of the observable Universe. Finally, the third problem asks why the Universe is close to being spatially flat. With the introduction of the Inflationary Model of cosmology in 1980s by Alan Guth, the three problems of the Big Bang cosmological model were solved. According to inflationary cosmology, the size of the Universe expanded exponentially to an extremely huge number (10^{60}) of its original size. This happened in a very short time from 10^{-35} to 10^{-32} s after the Big Bang. Collectively, the Big Bang model and Inflation Models of cosmology described the origin and expansion of the Universe.

By the mid-1990s, new observations led to new models of cosmology. The modern Standard Model of Cosmology, which is generally accepted among cosmologists, integrates the following theories, models, and concepts: a fixed background space-time, the General Theory of Relativity, Dark Matter, Dark Energy, initial conditions at Big Bang (best described by Inflationary Models), and the Standard Model of particle physics. Although the Standard Model of Cosmology has its own outstanding problems such as Dark Matter and Dark Energy, and issues with inflation, yet it explains all the observations.

Today cosmologists work with some of the most intriguing and yet fundamental questions such as:

- What is the overall shape and size of the Universe?
- Why did the Universe start in an improbable state?
- Did inflation happen?
- Is there a multiverse?
- What is the ultimate fate of the Universe?
- What is the true nature of Dark Energy?
- Can physics describe what was happening before the Big Bang?
- What happens to the notions of space and time before the Big Bang?

Currently there is no single theory that successfully answers all of the above questions. Beyond the Standard Model of cosmology, speculative theories in quantum gravity are used to research solutions for the problem of the selection of initial conditions of the Universe.

The purpose of this book is to present and explain the following unique topics in modern cosmology:

- A novel approach to uncover the dark faces of the Standard Model of cosmology.
- The possibility that Dark Energy and Dark Matter are manifestations of the inhomogeneous geometry of our Universe.
- On the history of cosmological model building and the general architecture of cosmological models.
- Illustrations of the large-scale structure of the Universe.
- A new perspective on the classical static Einstein Cosmos.
- Global properties of World Models including their topology.
- The arrow of time in a Universe with a positive cosmological constant Λ.

- Exploring the consequences of a fundamental cosmological constant Λ for our Universe.
- Exploring why the current observed acceleration of the Universe may not be its final destiny.
- Demonstrating that nature forbids the existence of a pure cosmological constant.
- Our current understanding of the long term (in time scales that greatly exceed the current age of the Universe) future of the Universe.
- The long-term fate and eventual destruction of the astrophysical objects that populate the Universe – including clusters, galaxies, stars, planets, and black holes.
- All evidence of the Big Bang including the Cosmic Microwave Background (CMB), and of the existence of other galaxies outside our own will disappear in about 100 billion years.

The material is presented in a layperson-friendly language followed by additional technical sections that explain the basic equations and principles. This feature is very attractive to readers who want to learn more about the theories involved beyond the basic description.

Each chapter is self-contained. Chapter 3 and Appendix A are related and collectively describe the future of the Universe.

Chapter 1 discusses a new perspective on the concept of Dark Matter and Dark Energy. The conventional view on Dark Matter is that these are particles that respond to the gravitational force, but they do not respond to strong, weak, and electromagnetic forces. We simply just have not discovered them yet. Whether they will be produced in the near future with LHC experiments at CERN, or detected with other experiments such as the Cryogenic Dark Matter Search (CDMS), remains to be seen. The nature of Dark Energy, however, is more mysterious than that of Dark Matter. It is believed to be the vacuum energy with negative pressure in its simplest form that causes the Universe to accelerate. It accounts for about 75% of the matter/energy of the Universe.

Chapter 1 argues that the Standard Model of cosmology may be just too simple, since it assumes that the Universe as a whole can be described by the homogeneous solutions of Einstein's equations. The chapter recalls the historical development of the Standard Model of cosmology and illustrates modern developments of our understanding of the large-scale structure of the Universe. It explains the shortcomings of the Standard Model and develops a more general model of cosmology that takes into account the inhomogeneities in the matter distribution, but also in the geometry of space-time. The tight relations between this geometry and global properties of our Universe, as implied by a full application of the General Theory of Relativity, provide possible solutions to the Dark Energy and Dark Matter problems. In this new framework, the classical Universe model, favored by Einstein, is put into perspective and, using this example, the limits and possible generalizations of the Standard Cosmological Principle are discussed.

Chapter 2 investigates the Arrow of Time in a Universe with a positive cosmological constant Λ. On the observed acceleration of our Universe, the analysis predicts that this acceleration is a temporary bleep and not our final destiny due to a fundamental scale Λ in nature. The chapter concludes that nature forbids the existence of a pure cosmological constant. This prediction can be tested by the combined observations from SN1a, large-scale structure LSS and, CMB, from existing or upcoming experiments such as Planck, Supernova Acceleration Probe (SNAP), and Laser Interferometer Space Antenna (LISA), will soon be able to pin down whether the equation of state of Dark Energy is a constant or if it evolves with time.

Chapter 3 covers the long-term fate of the cosmos. The evolution of planets, stars, galaxies, and the Universe itself over time scales that greatly exceed the current age of the Universe are addressed. The chapter follows the long-term development of stars and the stellar remnants (the neutron stars, white dwarfs, and brown dwarfs) remaining after the end of stellar evolution. Five ages of the Universe are summarized, namely: Primordial Era, Stelliferous Era, Degenerate Era, Black Hole Era, and Dark Era. The appendix at the end of the chapter outlines some of the basic equations that describe the astrophysical processes related to the future history of the Universe. It covers the Universe as a whole, galaxies, stars, planets, and black holes.

Appendix A demonstrates that in about 100 billion years (about ten times older than the current age of the Universe), all evidence of the Big Bang and of the existence of other galaxies outside our own will disappear.

I am grateful to the authors of each chapter, Professor Fred C. Adams, Professor Thomas Buchert, and Professor Laura Mersini-Houghton, for co-authoring the book and also for their patience throughout the book publication process.

I am indebted to Professor Lawrence M. Krauss for allowing me to include the article on "The Return of a Static Universe and the End of Cosmology" as an appendix in the book. I would like to extend my thanks to the other two members of the advisory council for the book: Professor Mark Trodden, for his suggestions and advice, and Professor David Finkelstein. I would also like to thank the staff of Springer, especially Jeanine Burks, for making this project happen.

Multiversal Journeys
March 2011

Farzad Nekoogar

Contents

1 **Dark Energy and Dark Matter Hidden in the Geometry of Space?: The Dawn of a New Paradigm in Cosmology** 1
   Thomas Buchert

2 **The Arrow of Time In a Universe with a Positive Cosmological Constant Λ** 51
   Laura Mersini-Houghton

3 **The Future History of the Universe** 71
   Fred C. Adams

Appendix A   The Return of a Static Universe and the End of Cosmology 119
   Lawrence M. Krauss and Robert J. Scherrer

Glossary 125

About the Authors 131

Index 135

Chapter 1
Dark Energy and Dark Matter Hidden in the Geometry of Space?

The Dawn of a New Paradigm in Cosmology

Thomas Buchert

1.1 The Dawn of the Standard Paradigm of Cosmology

Cosmology touches on the oldest questions of mankind. Is the Universe in which we live finite or infinite? What is the fate of the Universe? Can we predict it? Has it a beginning and/or an end? What are space and time? What is the structure of space and time, are they continuous or are there only discrete portions of space and time? Do space and time make separately sense in a unified description of a space–time? How many dimensions do we have to consider to describe it? Is there just one or are there many disconnected Universes, a Multiverse? Can we, in principle, see all of the Universe, or can we at least understand its overall architecture? How is it structured? Is there a large region that is typical for the whole Universe, so that we can learn all by just looking at this large region?

This list of questions can be certainly continued, but how do we start to gain more understanding? One approach is to consider the history of thinking about these questions. We may look at an important epoch in the history of what can be named "cosmology as a scientific discipline". This epoch is characterized first by constructions of valuable mathematical models of the cosmos, but equally important is the confrontation of these models with observations [13] (Fig. 1.1).

T. Buchert (✉)
Université Lyon 1, Centre de Recherche Astrophysique de Lyon, CNRS UMR 5574,
9 avenue Charles André, F-69230 Saint-Genis-Laval, France
e-mail: buchert@obs.univ-lyon1.fr

Un missionnaire du moyen âge raconte qu'il avait trouvé le point
où le ciel et la Terre se touchent...

Fig. 1.1 The flatness conjecture and the question of an infinite or a finite space reappear in the history of understanding the structure of our world. Riemann's geometry and Einstein's theory of gravitation have proposed solutions in terms of the possibility of curved finite space-forms without boundary. Still, contemporary cosmology rests on the assumption of an infinite space without curvature. This article explains a recently opened research field aiming to explain the Dark Energy and Dark Matter problems by generalizing the cosmological model in the spirit of this old question and its proposed answers [©Springer]

1 Dark Energy and Dark Matter Hidden in the Geometry of Space?

Fig. 1.2 Alexander Alexandrowitsch Friedmann (1888–1925) [©1993 CUP, reprinted with the permission of Cambridge University Press], Vesto Melvin Slipher (1875–1969) [©Springer], and Edwin Powell Hubble (1889–1953) [©Springer]

1.1.1 First Serious Encounter of Theoretical Cosmology with Observations

As Vesto Slipher in the early 1920s went out to measure redshifts of a handful of nebulae seen in the Milky Way, nowadays deciphered as other galaxies millions of lightyears away from our Milky Way, he would not know that he was laying the foundations for a paradigm in cosmology that survived until today.

We are in the time of Einstein's grand theory of general relativity that was fresh, and solutions of his theory were proposed by Alexander Friedmann, Abbé George Lemaître, Sir Arthur Eddington, Albert Einstein himself and Willem de Sitter. These first solutions were, of course, simple: they assumed the cosmos to be filled by a homogeneous-isotropic distribution of matter, i.e., the same at all places and in all directions of space (Fig. 1.2).

Edwin Powell Hubble added the decisive link to observational cosmology: Slipher's nebulae (by then about 50 measured) were distributed homogeneously and there was also no preferred direction in his distribution, so that – most naturally – the homogeneous-isotropic models that gave life to Einstein's equations were the ones that could stand for these observations. Since those nebulae were redshifted, which is interpreted as a recession movement away from us, the selection was clear: those models (out of a wealth of possible models, see Fig. 1.3) that describe an expanding space are the ones that correspond to what these measurements had to say.

1.1.2 Einstein's Cosmos: A Model That Has Been Abandoned

All this was so convincing to most of the researchers at the time that other ideas were left behind. For example, Einstein himself advanced a simple, also

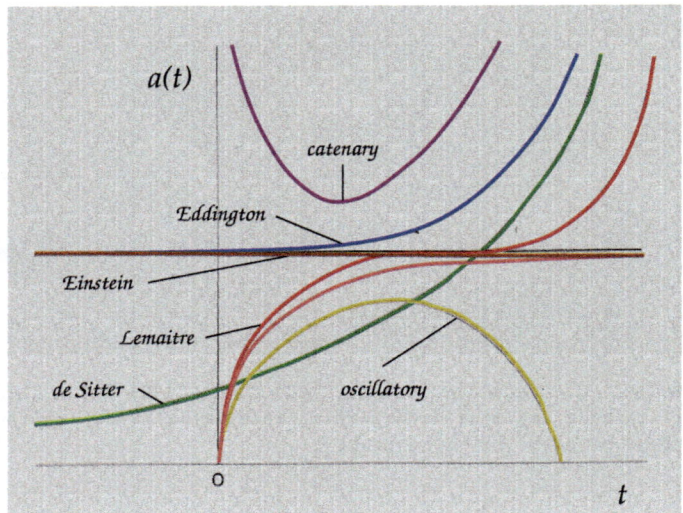

Fig. 1.3 The time-evolution of the scale-factor $a(t)$, measuring the separation of distances in the Universe, is shown, exemplifying the wealth of possible homogeneous-isotropic cosmological models. Standing out are: the Eddington model, which evolves out of the static Einstein cosmos and approaches the exponential expansion of the de Sitter model. The Lemaître models start out from a singularity (hence, the notion "Big Bang"), either having an inflection point (gradually entering the expansion phase), or oscillating back into a "Big Crunch". Models with positive constant curvature and no cosmological constant also oscillate back into a "Big Crunch", while models with no cosmological constant and negative curvature cross the static line and expand forever (not drawn). Among these is the Einstein-de Sitter model that assumes a flat space. There are also so-called catenary models that are "pulsating", i.e., they contract to a minimum and expand thereafter

homogeneous, model that was static, i.e., it did not change in time. To realize these thoughts of a space form that does not involve global evolution, he had to introduce a cosmological constant into his equations that would suitably balance the gravitational attraction of matter, so that the model has a chance to keep its shape in time without collapsing or expanding. Without this cosmological constant, space would be decelerating, i.e., if we start with an expanding space, then matter would slow down the expansion due to the attractive nature of gravity.

Since observations showed an expansion of the Universe, Einstein's cosmos was abandoned, and it did not help that Sir Arthur Eddington strongly argued for Einstein's model: he studied the instability of the Einstein cosmos and proposed later, based on a work by Lemaître on expanding universe models [17], to generalize Einstein's model as one that expands out of an Einstein cosmos and so does not have a "Big Bang". The latter could occur if we extrapolate an expanding model backward in time. A model that expands out of a static model would not have to be extrapolated back in time all the way to a state in which the contents of the Universe would be extremely compressed. In a sense, this backward extrapolation of the cosmos could be seen like a collapse of a star to always denser and denser

1 Dark Energy and Dark Matter Hidden in the Geometry of Space?

Fig. 1.4 *Left*: Albert Einstein (1879–1955) and Sir Arthur Stanley Eddington (1882–1944); [©Springer]. *Right*: Abbé Georges Lemaître (1894–1966) [©Springer]

structures until we reach a state of a "Black Hole". People were in favor of this extrapolation, since it allowed explaining the synthesis of nuclei as time proceeds and finally liberating radiation that no longer scatters with electrons in a dense medium, but – as atoms recombine – can propagate freely into space. The epoch when radiation is thus able to travel and finally reach us today, and this is the common view, can be analyzed in the maps of the Cosmic Microwave Background radiation. This background of radiation, which is about 2.7 K cold and that we can measure today, is thought to be the result of a hot state in the past: during the expansion of the cosmos this radiation was redshifted and thus has cooled down to this (almost absolutely cold) state today.

1.1.3 Eddington and Lemaître: A Spherical Finite Space or a Flat Infinite Space?

Lemaître published a paper following Eddington's view that described expanding models, however, where the initial state was still taken to be the static Einstein cosmos, not a "Big Bang". Later, he interpreted his universe model as *"the cosmic egg exploding at the moment of the creation"*, and Sir Fred Hoyle, although a critic of this idea, is responsible for the name "Big Bang". Eddington, and also Einstein disliked the idea. A cosmos emerging from a static state also shares, in addition to avoid the singularity problem, the beauty that guided Einstein through all his adventures, the beauty of a spherical space form, discovered by the brilliant mathematician Bernhard Riemann. The spherical space is positively curved and therefore finite in the same way as the surface of the Earth has a positive curvature and a finite extent. Here, however, we are in three dimensions, but the analogy prevails. One calls this a closed space without any boundary: we could move around in our Universe without hitting a border and still its extent is finite (Fig. 1.4).

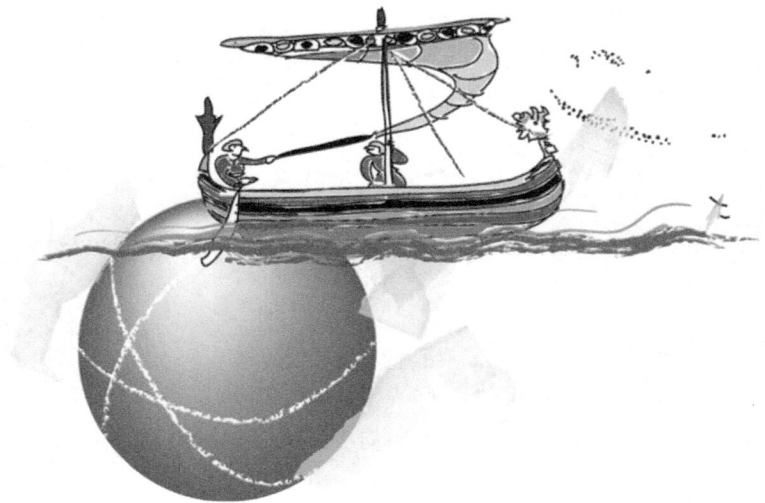

Fig. 1.5 A spherical Universe allows one to travel around without hitting a boundary. This is similar to travel on the surface of the Earth, which has a finite surface and, nevertheless, a traveller will never hit a boundary or "drop out of space". The surface of the Earth is the space for the traveller. In the Universe, we have to imagine one dimension more [Credit: Mauro Carfora]

Such a space form solves the problem that bothered many people before. For example, Giordano Bruno favored an infinite Universe, because if it were finite then one would hit a boundary somewhere and drop "out of space". In the spherical space this problem does not arise. There is no "outside of space" and we could in principle move to all points in that space and would eventually return to our starting point, see Fig. 1.5. Moreover, if the model is really static, the notion of time is globally unimportant, and the Universe itself inherits global non-changing properties.

Lemaître advocated an infinitely large, spatially flat cosmos that starts out with a "Big Bang", slows down and later would even accelerate, if he also included Einstein's cosmological constant. Looking just at the solution chosen as the cosmological model, a mathematician would have excluded the solutions with singularities (a "Big Bang"), but historical facts were already set out. The standard model was born, and the model of Lemaître is the one that nowadays is employed to interpret observations and brought into accord with them. It is now coined as "Concordance model of Cosmology".

1.2 The Invisible Universe: An Old Model with a Modern Dark Face

Gone are the times when people wrote long letters to each other before they even dared to publish their results. As there was a handful of galaxies, there also was only a handful of cosmologists. Cosmology was in its childhood and the new theory of

1 Dark Energy and Dark Matter Hidden in the Geometry of Space?

the Universe was young and had to be explored. Today, we instead have catalogues of millions of galaxies that map out the Universe as we collect the redshifts of nearby and very distant galaxies into what we call "the large-scale structure of the Universe".

1.2.1 The Large-Scale Structure of the Universe

Slipher's sample of "nebulae" was very sparse. It covered all the sky and so these around 50 galaxies could not trace the detailed structures around us. This sample appeared homogeneous and isotropic and so it did not hint to a wildly structured Universe. This new picture of the world extended smoothly the picture that Einstein and his collegues still had in the 1920s: the Universe was identified with our own galaxy, the Milky Way, and models were in fact conceived for the Milky Way Universe. That these nebulae were other galaxies opened the view on the world dramatically. At least these galaxies seemed to float around in an ordered fashion, so that one could speak of a homogeneous density distribution of matter. A much more dramatic change of this picture was set out in the 1980s. The Center for Astrophysics Survey revealed the first "large-scale structure": a "stick man" appeared in the sky, made of a bridge between the Virgo and Coma rich galaxy clusters and big walls of galaxies that surrounded two voids besides this bridge, see Fig. 1.6.

1.2.2 The Construction of a Galaxy Map

Let us look into more details on how such a map of the Universe is constructed. Observing a galaxy determines its two coordinates on the sky. A third coordinate is provided by the redshift, interpreted as recession velocity of the galaxy as it is partaking in the global expansion of the Universe, including smaller velocities relative to the global expansion.

If we map these redshifts to a distance of the galaxy, we already need a model that relates the observed redshift to a distance scale in space. The standard model does provide this: given the model we can determine the distance that light would travel in this model. The proper motions of galaxies with respect to the global expansion introduce just a small mistake: galaxies in rich clusters form "fingers" that point to the observer, which can be seen and are explained in Fig. 1.6.

The so-constructed three-dimensional maps of the Universe do not show a homogeneous and isotropic distribution, but rather galaxies are agglomerated into clusters of hundreds or even thousands of galaxies. Rich clusters form "knots" in a large honeycomb structure whose cells surround regions that are almost devoid of galaxies. In a much larger survey, the Sloan Digital Sky Survey, see Fig. 1.7, we even see walls extending over a billion lightyears.

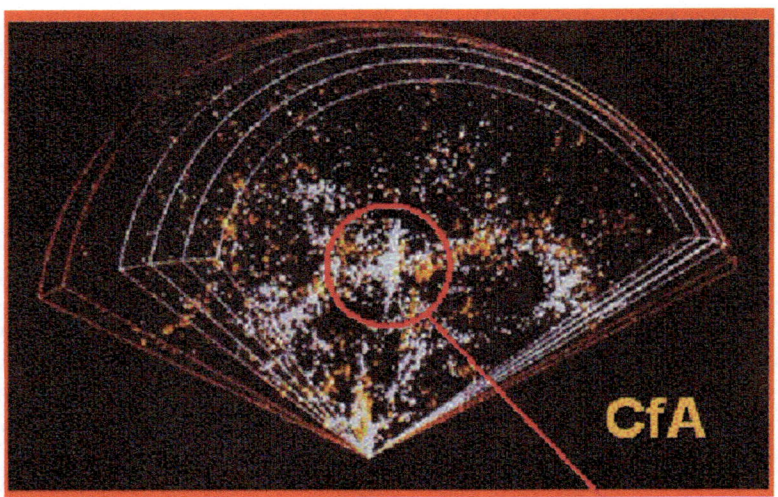

Fig. 1.6 Redshift slices of the CfA Survey (Center for Astrophysics Survey) galaxy data around the galactic equatorial plane. This survey shows a map of galaxies as seen around us (the bottom of the wedge) in the equatorial plane by collecting the thin slices together. It reaches out to ≈ 150 Mpc/h, where h is a factor of the order of 1 depending on the Hubble constant measuring the expansion speed, and 1 Mpc is about 3 millions of lightyears. Each slice contains $\approx 1{,}000$ galaxies. This catalogue shows the highly inhomogeneous structures: galaxies form thin sheets (the most pronounced sheet pattern is called the "Great Wall"), surrounding rarely populated regions (so-called "voids" of typical extent of ≈ 50 Mpc/h). A three-dimensional cellular structure results, two-dimensional cell walls represent superclusters of galaxies, one-dimensional string like structures ("filaments") connect cell knots which consist of rich clusters of galaxies like the Coma cluster in the center of the survey. The "stickman" pattern around Coma is due to "fingers" which point towards the observer. These are the result of the relative motions of galaxies with respect to the global expansion. They are visible by mapping redshifts of galaxies into distances using the standard model that describes the global expansion only [Credit: John P. Huchra, CfA]; see [15]

These maps of the Universe show that the assumptions of homogeneity and isotropy in the standard cosmological model, that are now also entering as a prior into the distance measurement, have to be rethought: the matter distribution is obviously not homogeneous and also the counting of galaxies into different directions would be different. It appears that the idea that the Universe should look the same around every observer, could only be realized if we look at very large scales (billions of lightyears). This scale is certainly beyond these maps of the regional Universe. Moreover, we can expect this only in some averaged or statistical sense. We shall come back to this question and the issue of looking at the averages of a matter distribution later.

1 Dark Energy and Dark Matter Hidden in the Geometry of Space? 9

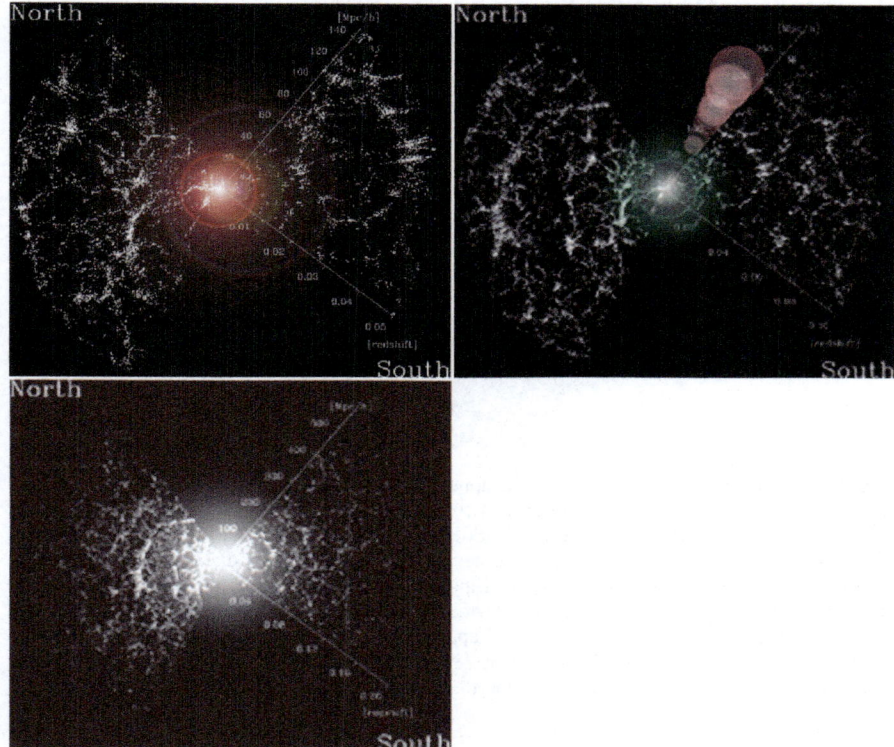

Fig. 1.7 Redshift slices of SDSS (Sloan Digital Sky Survey) galaxy data around the galactic equatorial plane. The redshift limits and the thickness of the planes are: *Upper*: $z < 0.05$, $10/h$ Mpc; *Middle*: $z < 0.1$, $15/h$ Mpc; *Lower*: $z < 0.2$, $20/h$ Mpc. In these pictures we see that structures are organized into almost empty regions that are surrounded by walls of galaxies. Most of the volume of the present-day Universe is in empty space. (see [16].)

1.2.3 The Dark Sectors of the Standard Model

Today, thousands of scientists are contributing to the development of cosmology, many of them may not even understand Einstein's general theory of relativity and they may even work in completely different research fields. Huge astronomical projects have been and are going to be launched, are analyzed and realized by huge communities of engineers.

However, all of them suffer from one big problem: the model they use is the old model, but its physical energy content seems to have disappeared (see Fig. 1.8).

The matter seen by its light emission together with ordinary matter that does not shine seems to make up a small fraction of the sources for Einstein's equations

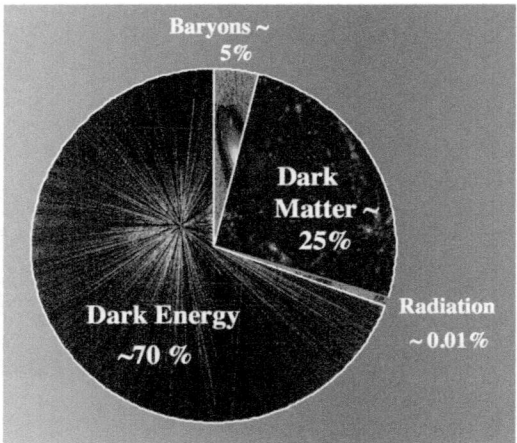

Fig. 1.8 The invisible part of the Universe appears as a parametrization of physical ignorance in the standard model of cosmology. A physical cosmology must explain the two dark sectors: Dark Matter acting gravitationally like matter, Dark Energy acting anti-gravitationally like a repulsive force. Known matter (baryons and massive neutrinos) and energies (radiation) make up a small fraction of the sources necessary to satisfy Einstein's equations for a flat spatial geometry. Note that, in the theory of relativity, matter is a form of energy by the famous relation $E = mc^2$ with the speed of light c. The numbers given are approximate. More precisely, today Dark Energy is thought to make up 73%, Dark Matter about 23%, and baryonic (known) matter about 4%. Note that only about a decade ago this "camembert" looked very different, since the presence of Dark Energy was not yet discovered

(at most 5%), if this model is used to interpret the data. In place of ordinary matter a growing fraction of the content of the Universe seems to be "dark". In the course of an enormous improvement of observational data during the last 30 years the "camembert" of Fig. 1.8 has been more and more "eaten up" by dark components of sometimes odd character.

People postulate particles, so-called *Dark Matter*, not yet detected in experiments, that do not radiate and that would interact only gravitationally and very weakly with ordinary matter. There are essentially three research fields that touch upon three questions: (a) how do we know that there must be Dark Matter? (b) how could we find Dark Matter particles in experiments? and (c) what are the candidates in a particle physics theory? The answer to the first question is thought to be given: Dark Matter is missing gravitationally, if the dynamics of a galaxy, a cluster of galaxies or the large-scale structure is analyzed and confronted with the known matter ingredients. For galaxies, for example, the measurement of rotation speeds as a function of distance from the center reveals that the galaxy would fly apart without Dark Matter: a large Dark Matter halo around the galaxy is needed to hold it together. It should, however, be said that all these estimates are model-dependent and the models used are rather simple. There are efforts to build more sophisticated, general–relativistic models for galaxies, and those studies show that it may not be so simple and the need for Dark Matter may be not so obvious. In order to answer the second question many experiments have been designed and are operating now. One example is the mountain at Gran Sasso in Italy that is used to build underground detectors for the unknown dark particles. The idea is to filter the cosmic particles while they scatter through the matter of the mountain and mainly the weakly interacting ones, so-called WIMPS (Weakly Interacting Massive Particles), would remain after a long travel through the mountain. These are often smart and difficult experiments, where the instruments have to be cooled down to extremely low temperatures, and measurements have to be done during long periods of time in order to obtain reliable statistical information. Single events that are discovered have to come up more often and they have to be checked by other experiments in the world. (A single event may be just the result of a truck carrying radioactive material that travels without permission through the tunnel of Gran Sasso.) The third question is mainly a question for particle physics theory: the standard model of particle physics does not predict the existence of such particles. More general theories are developed in which known particles have "supersymmetric partners". A "minisuperspace" is constructed to narrow down the number of possible candidates, and observational limits from experiments paint forbidden and allowed regions in this space. Also, these regions are model-dependent, and they may vary by varying some parameters. It may also be that Dark Matter is formed by known matter. We already know that there are many non-shining objects flying around, e.g., of the size of Jupiter, so-called MACHOS (Massive Astrophysical Compact Halo Objects). Observational experiments are conducted to search for these and to quantify them better. Also, there may be swarms of mini-Black Holes (possibly already created in the Early Universe). Black Holes that may generically sit at the galactic centers are thought not to largely influence the gravitational budget of a galaxy halo, but here general–relativistic models have not yet been studied thoroughly to exclude this possibility. Overall, as long as no Dark Matter particle is detected (and even if it is detected and the amount not being sufficient), the first question concerning the improvement of models is certainly one that must not be abandoned: maybe we do not need Dark Matter in the first place?, a question that can still be posed.

Furthermore, we seem to need *Dark Energy*, maybe in the form of exotic fluids; a postulated field called *Quintessence* (a notion invented in the Middle Ages) that, in addition to ordinary matter, Dark Matter, radiation and an eventual homogeneous curvature would add a fifth element. This fluid should act in a repulsive way, that is, it should have a negative pressure that counteracts gravity. Some physicists think that there are "phantoms", "ghosts", names invented for fields that violate normal physical properties of known matter and that also go beyond the standard theory of particle physics. The outskirts of Geneva in Switzerland reside on rings of an accelerator, the LHC (Large Hadron Collider) that is just now starting to probe energies that eventually unveil the dark mysteries carried by yet unknown fundamental fields.

Many respected scientists are serious about these dark components, but others doubt that these exotic particles and fluids may exist at all. Looking at these industry-scale experimental efforts it is surprising that theory did not really catch up. An astronomer using today's Cosmic Microwave Background explorers, measuring the relic temperature background of the Universe, could easily be understood by Friedmann, as far as the cosmological model is concerned. All observational and experimental results are first packed as a burden on the standard model.

1.3 A Serious Problem with the Standard Model

We recall: the standard model of cosmology, going back to Lemaître, is a homogeneous and isotropic solution of Einstein's theory of gravitation with a cosmological constant. It is nowadays coined as the standard "concordance model" of cosmology, if its parameters (see the appendix for more details) are determined to fit observational data. These latter are gained on the basis of huge observational and experimental efforts. It turns out that known matter and energies would only provide a fraction of a few percent of the sources that are needed to satisfy the equations of the model. While the status of observational cosmology has dramatically improved, the theoretical part concerning the basis of the cosmological model has been the same for about 80 years. Observations determine at high precision the initial conditions of this model.

Already Isaac Newton in his *Principia* [18] warned that physics is not concerned with initial conditions, but rather with the understanding of the laws of nature (Fig. 1.9). We are exactly in this situation: given the model and confronting it with observations to determine its parameters, we must postulate that there are missing "dark" sources. The standard model of cosmology is, thus, not a "physical cosmology", since it does not explain the physical nature of these sources. These could be particles, i.e., fundamental fields, that we have not yet seen in experiments. There is, however, another research field that even abandons Einstein's theory of gravitation.

Fig. 1.9 Sir Isaac Newton (1643–1727) [ⓒSpringer]

1.3.1 Is Einstein's Theory of Gravitation Wrong?

There are theoretical efforts that fly off into a completely different direction: many scientists think that, in case there are no exotic fluids and particles, then the laws of gravitation must be wrong. They invent theories other than Einstein gravity in order to see whether, for example, higher-dimensional spaces may eventually cause an effect similar to that of the dark sources. Examples include "braneworld cosmologies" which assumes we are living on a "brane" (the usual three-dimensional space), but that gravity can live in more than three dimensions. Observational data are then exposed to these theories of even more exotic nature than the unknown fields that they are supposed to describe. Einstein's theory of general relativity is seriously put into doubt. Clearly, they say, we need more general theories of gravitation, as the standard model, being a solution of Einstein's laws of gravitation, fails to explain the Universe, because it needs dark vehicles to interpret observations. There are, of course, other arguments and motivations in theoretical particle physics, along with the attempt to establish a quantum theory of gravitation, that all indicate the need for a generalization of Einstein's theory, but the above line of argument is shortsighted in the sense that the "dark problem" may be a classical one that could be resolved within the well-tested theory of Einstein. Regarding the systems where such tests have been conducted, e.g., on solar system scales, even Einstein's theory is already an extrapolation to cosmological scales. At any rate, a cosmological model built on a special solution of the general theory of relativity of Einstein should not be confused with the theory itself that, as we shall learn, has a much richer tone as the simple models of Friedmann and Lemaître, and it allows the construction of more general cosmological models.

1.3.2 The Architecture of Current Structure Formation Models

While the standard homogeneous-isotropic models are supposed to describe the global evolution of the Universe, locally there are structures. Several methods have been used to describe the evolution of these structures. One class of methods uses the fact that the homogeneous-isotropic models are unstable to inhomogeneous perturbations. These perturbations in turn are described in such a way that the overall distribution of matter, i.e., if averaged over some large scale, agrees with the homogeneous-isotropic models. One speaks of a *background space* and deviations off this background. These deviations can be large, which can be handled by going to higher-order perturbations, i.e., by taking into account more terms in the expansion around the background, or, by numerically simulating the full inhomogeneous system of gravitational equations. This latter is currently realized by numerically integrating the Newtonian system of gravitational equations.

Thus, the structure formation process is seen to evolve on a fixed stage: the geometry of space is, in the simplest case of a spatially flat background, Euclidean and remains so. Technically, a numerical simulation of structure is realized in a box

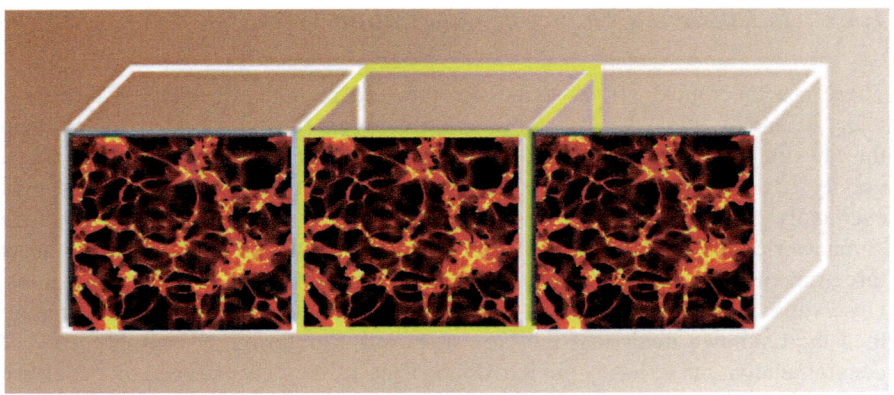

Fig. 1.10 Current simulations of structure formation in the Universe are based on a simple architecture: a box is introduced that follows the expansion of the standard model of cosmology. Structures are described to be periodic on the scale of the box that implies that the average over the matter distribution in the box always vanishes. As a consequence the global evolution of the universe model is not influenced by the structure formation process by construction. The Universe is seen as a replica of identical boxes, justified by the argument that the box should be large enough to describe typical properties of the matter distribution in a statistical sense. The cosmological principle demands that every observer would see the same statistical properties of the matter distribution, which allows extrapolation of the properties seen in a single box. The simulations are Newtonian, i.e., the geometry is Euclidean in space and remains so

with some large length, see Fig. 1.10 for the principle architecture, and Fig. 1.11 for a modern large simulation. The deviations from the background form structures and one guarantees during the simulation that the average of the distribution vanishes for all times, as in perturbation theory mentioned above. The point is that this architecture implies, by construction, that the universe model – containing now structures within the box – is still globally described by the background, i.e., the simple model of Lemaître. More generally, however, the structures influence the background evolution, as we shall explain below in detail. Let us finally remark that the above architecture is the only one possible, if structures are described in Newtonian theory, since this latter needs specification of boundary conditions at all times (here: periodic boundary conditions, see Fig. 1.10) to assure unique solutions. This will not be necessary, if structures are described within the general theory of relativity of Einstein.

The universe model based on Newtonian cosmology is thus a kaleidoscopic partitioning of the Universe into boxes. This model appears quite rigid in the framework of Einstein's theory. We shall be concerned later with the geometrical aspects of Einstein's theory showing that structures do not, on average, behave like the background model of Lemaître, but instead the background and the structures will interact. Also, a flat periodic structure cannot be implemented as in the Newtonian model. We may imagine the following picture: to require that the average model remains identical to a fixed background model for all times, is like a ship that

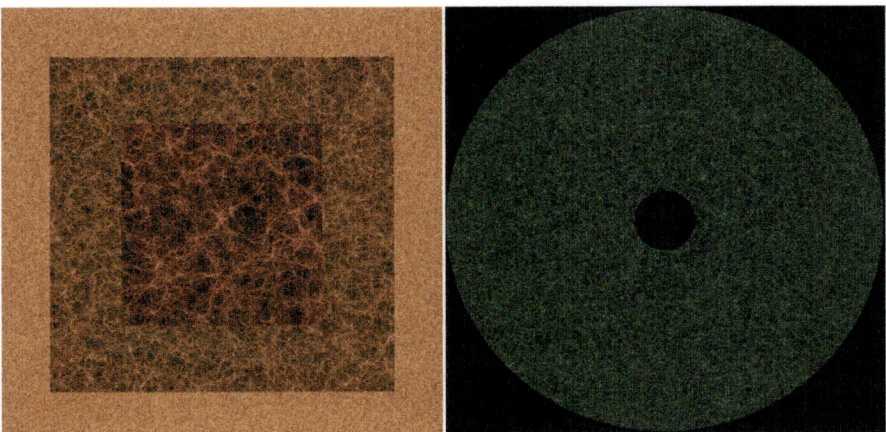

Fig. 1.11 *Left*: A multi-resolution view of the HORIZON simulation of a periodic universe model. The outer region corresponds to a view of the Universe on scales of 16,000 Mpc/h: it is generated by unfolding the simulation while cutting a slice obliquely through the cube in order to preserve the continuity of the field (thanks to the periodicity). The intermediate region corresponds to a slice of 2000 Mpc/h, while the inner region is at the original resolution of the initial conditions. *Right*: A redshift slice through the HORIZON simulation of a periodic universe model [Credit: HORIZON project 2009]

is kept from floating away by an anchor, even if the wind blows strongly. Einstein's theory allows the anchor to be lifted into the ship so that it can (and will) float away: on average the universe model drifts away from the background of Lemaître. We shall later explain this further.

1.3.3 Why Is the Standard Model Too Simple?

Most recently, there is a growing number of researchers that neither claim the existence of exotic dark sources, nor do they want to abandon the theory of Einstein for the description of our space–time. The standard model is simple, since it assumes that the Universe as a whole can be described by the homogeneous solutions of Einstein's equations discussed above. The Universe, however, is no longer made up of the few nebulae of Vesto Slipher: there are enormous structures in the Universe that do no longer justify using a simple distribution of matter. This is obvious from Fig. 1.7, but why do cosmologists keep nevertheless up with this model? The answer is, although often not spelled out explicitly: the Universe is inhomogeneous, but if these structures are "averaged over", the so-obtained homogeneous distribution can be described by a homogeneous solution of Einstein's equations. This is the conjecture underlying the standard model.

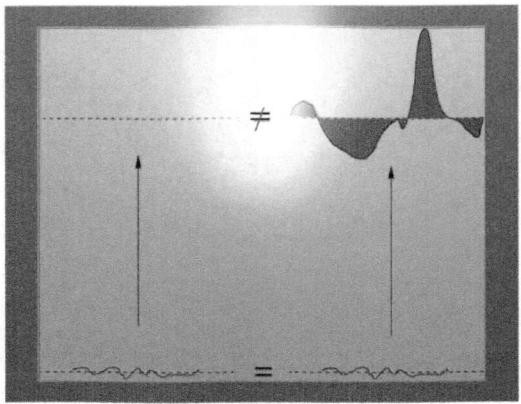

Fig. 1.12 We compare two different views on the structures in the Universe: on the left we take the average over some tiny matter fluctuations and use those as starting values for a homogeneous solution of Einstein's equations (the standard model); on the right, more realistically, we evolve the tiny matter fluctuations with the full Einstein equations and, of course, today we shall find the large structures seen for example in Fig. 1.7. We may then average over this inhomogeneous distribution. The crux is that the resulting average field is not the same as that obtained by evolving from the beginning a homogeneous distribution. One says that the two processes (**a**) time-evolution and (**b**) averaging are "non-commuting". The standard model, however, if it were to describe the averaged Universe, must assume that the average on the right is the same as that on the left

This sounds overall reasonable, but there is a serious misconception: it is not at all clear that an averaged realistic model with structures would agree with the standard model, a problem that has been pointed out earlier and emphasized especially by George Ellis of the University of Cape Town in South Africa already in the 1980s (see [12, 13] and references therein), and thereafter followed with concrete models by Toshifumi Futamase of the University of Sendai, Japan, Masumi Kasai of the University of Hirosaki, Japan, and Mauro Carfora of the University of Pavia in Italy. Indeed, it is not in general true as Fig. 1.12 explains.

We conclude that a universe model that describes the evolution of the average of the real structures does in general not evolve as the homogeneous solutions. Some people think that this may be a small correction, others already demonstrated that this difference is crucial if all consequences are studied. For example, already here the observational interpretation of data must change, since we determine the parameters of the model by comparing observations at very different epochs of the universal history: astronomers look back in time and they may analyze the Cosmic Microwave Background at a redshift of about 1,000, or they may analyze the galaxy maps that reach out only to redshifts of a few. Common comparisons of these results are based on an evolution model that connects the parameters at the different epochs. If these parameters evolve differently, such a comparison delivers different conclusions.

1.4 A New Cosmological Model

It is clear that the above conjecture of commutation underlying the standard model of cosmology must hold at least approximately, before it makes even sense to apply this model to describe our inhomogeneous Universe. We may, however, assume that the difference explained in Fig. 1.12 is small, and this is believed by many researchers, so that we could still use the standard model as a good approximation. But, this has to be shown! An attempt to take structures in the Universe into account and to really do the average over a complicated distribution is necessary.

Given this insight of a principal difference between the standard homogeneous model and a model that is construed as an average over structures, we are going to aim at quantifying this insight in order to generalize the cosmological model, while remaining within the theory of general relativity. Before we explain how we can do this, it is worthwhile to recall the basic traits of Einstein's theory.

1.4.1 Back to the Roots: Einstein's Theory of General Relativity

It is important to spell out the main qualitative differences between Einstein's theory of gravitation and the former Newtonian theory. We have learned that the evolution of structures is described within the Newtonian theory on a fixed background. This background is also a solution of Einstein's theory, but it is not for the evolution model including structures. In principle we could aim at building a simulation model within general relativity, and this has been done for special configurations like "Black Holes". For a universe model, however, this is a difficult task. Also finding sufficiently general solutions of Einstein's theory that are inhomogeneous is very difficult. Is there a strategy to proceed in a simpler way by just relaxing some restricting assumptions of the standard cosmology in order to capture important aspects of Einstein's theory? (Fig. 1.13).

Fig. 1.13 Johann Carl Friedrich Gauss (1777–1855), Georg Friedrich Bernhard Riemann (1826–1866), and Albert Einstein (1879–1955) [all images ©Springer]

Let us explain what these important aspects are by looking at the structure of Einstein's equations. Symbolically, these equations are written as

$$\mathbf{G} = \kappa \mathbf{T},$$

where **G** describes, in geometrical terms, the structure of a four-dimensional space–time, **T** the matter and energy sources, and κ is the gravitational coupling constant. Simplifying the geometrical side, as is done in the standard model, we obtain simpler equations that balance the geometrical terms on the left and the sources on the right (we write down these simpler equations in the appendix). We discussed that this simplification to a homogeneous geometry on the left would generate the need for unknown homogeneous sources on the right to satisfy the equations. It is obvious that, if the sources are in reality inhomogeneous, i.e., display structures, also the geometry on the left must be inhomogeneous. What happens for the sources on the right must happen correspondingly in the geometry on the left. Einstein's equations couple the geometry to the sources, and these sources evolve in the space–time dictated by its geometrical structure. This is far different from describing structure formation on a geometrical stage that is not changing like in the Newtonian theory. One says that the geometry of space–time is a dynamical variable; it is not in Newtonian theory, since the fixed space–time is absolute and does not "talk" to the structures. If it does, there are (eventually complicated) interactions between geometry and matter, and Einstein's theory is indeed formally quite involved.

1.4.2 Construction of a Model That Does the Average Over Structure

As illustrated in Fig. 1.14, we are able to construct a more realistic model that does the average over an inhomogeneous distribution of matter on a dynamically evolving spatial geometry. We have to "average out" the inhomogeneities so that we obtain a homogeneous model for the matter sources that, and this is the crucial difference to a Newtonian model, then are evolving in a geometry that is changing. (Details on the refurnishing of the cosmological equations including structure in relativistic cosmology may be found in the appendix.)

Nevertheless, the resulting average model turns out to be quite simple: it obeys almost the same equations as the standard model, but with additional geometrical terms. These terms can play the role of Dark Energy and even Dark Matter, as it is schematically explained in Fig. 1.15.

1.4.3 Curvature of Space: An Important Player Is Rediscovered

Einstein himself would not be surprised at all: the geometry of our space–time is an important player. His theory is exactly based on an intimate interaction of

1 Dark Energy and Dark Matter Hidden in the Geometry of Space?

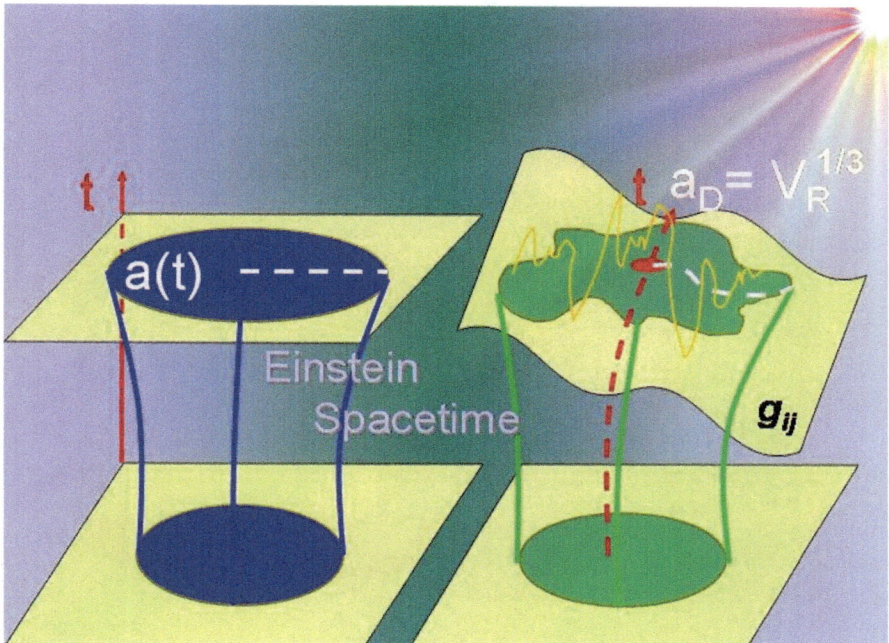

Fig. 1.14 The construction of a more realistic model is schematically drawn. The standard model, on the left, evolves a homogeneous distribution of matter. You may think of a sphere that evolves into a sphere with a radius $a(t)$ given by the homogeneous solution of Einstein's equations. On the right we appreciate that this sphere is distorted, if the changing geometry of space (caused by the formation of structure) is taken into account. In both cases we assume that the amount of matter inside the domain is the same at all times. If we, moreover, consider an inhomogeneous distribution on such a geometrically distorted domain, then we may look at the average over this distribution, but on the real geometry of space, not – as is usually done – on a flat absolute space that was the way to view structures introduced by Newton. The volume of the sphere on the left, $V \cong a^3(t)$, may then be replaced by the volume of the actual, curved domain, $V_R \cong a_D^3$. The non-restricted Einstein equations then determine the evolution of this realistic domain through the evolution of the inhomogeneous geometry of the space, and the resulting averaged model turns out to be different from the standard model

structures in the matter distribution and the geometrical properties of our world. Looking at the standard model in the case of flat space (the concordance model), it is so simple that it leaves the geometry unchanged. Structures are then described by simulations, as the one in Figs. 1.10 and 1.11, on a flat space in which distances are still measured with the elements of Euclid and Pythagoras. Nothing seems to have changed after the discovery of Einstein: Newton's theory is considered sufficient to describe gravitational interactions in our Universe. Even the standard model can be recovered as a solution of Newton's laws of gravitation without solving Einstein's equations. Einstein's theory is not even needed to realize the present-day models.

Curvature of space does, however, not only naively change the distribution of matter, as it is suggested in Fig. 1.16. The more realistic model shows that the

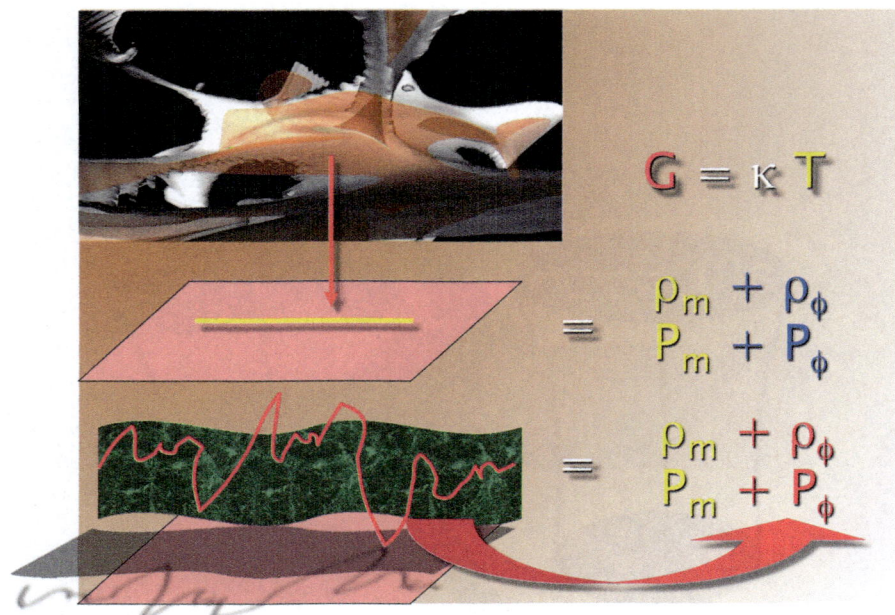

Fig. 1.15 The standard "concordance model" of cosmology (*upper*) idealizes space as being flat and the distribution of matter on this space being homogeneous. Einstein's theory relates the sources with the geometry of our space–time. In order to explain observational results, the sources that must be assumed to generate this simple geometry contain not only known sources (indexed with m), but also unknown, postulated dark fields (indexed with Φ). The more realistic model (*lower*) evolves geometry and matter unrestricted. After averaging, the result is a model that can be written as a standard model with additional geometrical terms. If these extra terms, that emerge from the inhomogeneous distribution, are put on the right, the more realistic model looks like a standard model, but at the same time mimics a source that must no longer be attributed to mysterious dark components, but that here naturally emerges from structure. The remaining, still an open question in present-day research is, whether this generated component is sufficient to explain the observations

converse also holds true: as soon as structures form, they also change the curvature of space. Again, Einstein would not be surprised, since the very meaning of his theory implies that the matter distribution "talks" with the geometrical structure of space–time, and if the structures evolve, then also the space–time geometry is changing. So, this mechanism is actually not new, but curvature is not taken into account at all in the standard descriptions. Let us look more closely at this mechanism in the form of an equation that does not exist in a Newtonian theory of gravitation. We may symbolically write this formula as (see the appendix for the concrete formulae):

$$\text{change of structure in matter } Q = \text{change of space curvature } R. \tag{1.1}$$

1 Dark Energy and Dark Matter Hidden in the Geometry of Space?

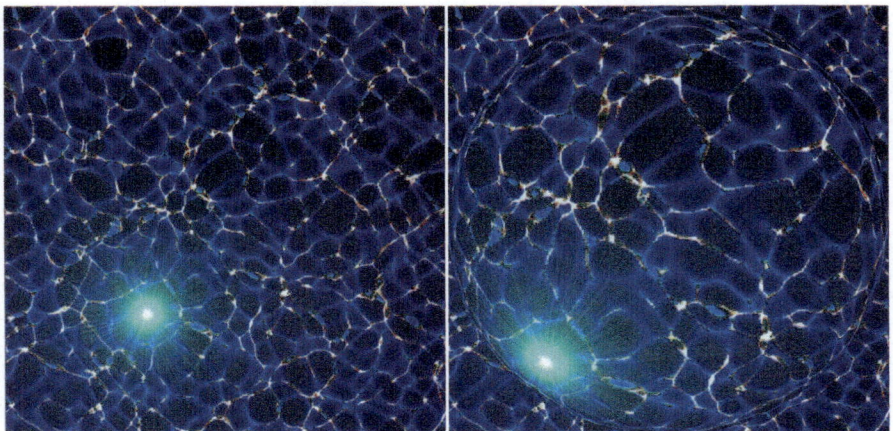

Fig. 1.16 Current simulations of structures in the Universe use the old Newtonian theory. Newton assumed the existence of absolute space and time, an unchanging stage on which the structures of a fluid evolve. Einstein's theory abandons these notions of space and time: the geometrical structure of the space–time is supposed to be identical to the gravitational field generated by the structures. This field can be curved and the geometry of our space–time is dynamical and interacts with the evolving matter distribution. This figure is an artistic illustration of the effect of curvature of space: a small curvature could lead to drastic changes of the matter distribution [Credit: [8], reproduced with permission ©ESO]

This formula is easy to interpret: as soon as structures in the Universe form (Q changes), they influence the geometry of space (R changes), and it is the same space in which these structures evolve. Moreover, there is a tight link between structures inhabiting the Universe and the way we observe those structures: light follows the geometry of space–time and we are observing structures by their emission of light. It is clear that our interpretation of the structure distribution then changes too. The relation (1.1) does not exist in the standard model: the curvature R stays the same all the time, while fluctuations Q are evolving on this unchanging geometry. In the more realistic model it is this link between structure evolution and the curvature of space that may create the illusion of Dark Energy and eventually also Dark Matter, as we explain now.

1.5 Dark Energy and Dark Matter: Are They Hidden in the Geometry of Space?

We recall: a more general cosmological model that takes into account the structures in the Universe can be written like the standard model with ordinary matter as source, but in addition we have some extra sources. These latter are not fundamental, i.e., they do not correspond to existing particles and fields, but they emerge from the (averaged out) inhomogeneities. They arise first on the left hand side of Einstein's equations $\mathbf{G} = \kappa \mathbf{T}$, but if we just keep the geometrical terms of the standard model on the left and interpret the extra geometrical terms as sources by putting them to the right hand side, we obtain model equations that look like standard expansion and acceleration laws, but here expansion and acceleration are caused by more terms than the usual matter source. Note that we can do this, since we do not change the equation that must be satisfied. Since these extra terms arise in general, it is tempting to associate them with the missing dark energies in the standard model. But: does this work?

1.5.1 The Acceleration of the Volume Expansion of Space

In the standard model, a recent acceleration of the expansion rate is needed to explain the apparent dimming of distant objects. Observations of distant supernovae that seem to be further away from us than a model without a recent acceleration would predict, led people to include Einstein's cosmological constant as a simple model for this repulsive fluid, dubbed Dark Energy, to explain this. The "coincidence" that this acceleration seems to happen around the time when the structures in the Universe develop strong deviations from homogeneity hints at a connection between this global effect of acceleration and the formation of structures. Note that we observe the dimming, but we do not "observe" the acceleration: only by assuming the standard model to hold, we need that the Universe accelerates. Care must be taken also in comparing the acceleration of the cosmological model and the "apparent acceleration". The latter we observe in the direction of incoming light, while the former is a spatial property. In the standard model both are identical, but in an inhomogeneous model there is the possibility that we observe a dimming of supernovae without an acceleration of the volume expansion of the model, a possibility that has been explicitly demonstrated for a spherically symmetric space–time.

We are trying to understand this acceleration within the more general model. For this purpose we take a closer look at another equation of the more general cosmological equations. This equation describes the acceleration of the volume expansion of our Universe:

$$\text{acceleration of volume expansion} = -\text{ density of matter} + \Lambda + Q. \quad (1.2)$$

The term Q encodes the inhomogeneities as above, it is often called "backreaction", since the presence of structures "reacts back" on the evolution of the standard model. The term Λ is Einstein's cosmological constant. Let us first look at the standard model: it assumes $Q = 0$ (no inhomogeneities or no structures). Then, if we would have only ordinary matter, there would be no acceleration (there is a negative sign in front of the density); matter slows down the expansion of the Universe due to the attractive nature of gravitation and so predicts a negative acceleration. Only if we include a positive Λ larger than the density (which is the simplest model for Dark Energy), we can explain an acceleration of the volume expansion of the Universe.

In the more realistic model this situation changes: a positive term Q could play the role of a cosmological constant (Dark Energy), a possibility that has been recently considered by many researchers comprising Rocky Kolb of the University of Chicago in the US, Sabino Matarrese of the University of Padova in Italy, and Syksy Räsänen of the University of Helsinki in Finland. Although this point of view would also solve the "coincidence problem" that the acceleration occurs around the time when the Universe becomes structured, the argument needs a rather strong Q larger than the density to achieve this. However, as stated above, we may not need an acceleration in this general model to explain the observed dimming of supernovae, which is the only reason to postulate Dark Energy to exist.

What gives an extra swing to the possibilities of the backreaction term, a negative Q could also mimic a contribution to matter (Dark Matter). Now, it turns out that a positive Q would change, via formula (1.1), the curvature of space to be negative, and a negative Q would produce a positive curvature. Dark Energy and Dark Matter could thus be associated with different geometrical space properties. Moreover, it can be shown that large, almost empty regions (to be seen in Fig. 1.7) would have a negative curvature, and the places where galaxies reside must have a positive curvature. This is compatible with what we need: Dark Matter is a component that seems to be missing within matter-dominated regions, and Dark Energy is a component that influences the large-scale expansion property of space. The argument is then the following: if we find acceleration of the volume expansion, then this volume must be dominated by almost empty regions, and indeed this is suggested by the observations of the distribution of structures (Fig. 1.7), but also by the simulation of structure formation (Figs. 1.11). In the course of time, more and more empty regions occur and matter is condensed into tiny fractions of the space volume (a model that exemplifies this very nicely has been recently given by Syksy Räsänen while working at the University of Geneva in Switzerland [20].)

Hence, an explanation of the dark components is at least suggested, if the model accelerates or not, in terms of the inhomogeneities in the curvature of space. Since the standard model neglects this curvature variation, it is "blind" in this respect and that maybe the reason why it has to carry additional dark sources.

1.5.2 An Expanding and Curved Space Emerging from an Empty Universe: A "Gedanken Experiment"

Let us hold in for a moment and imagine that we look at a portion of the Universe that is empty, i.e., there are no matter or energy sources. Einstein's equations are still valid, but they now describe the geometrical structure of an empty world (the vacuum). The empty four-dimensional space–time is, according to Einstein's theory, not necessarily a portion of the so-called *Minkowski space–time* with everywhere vanishing curvature. We have to be more precise here: Einstein's equations relate the sources to the so-called *Ricci curvature tensor*, which only describes part of the full curvature. This full curvature, encoded in the so-called *Riemann curvature tensor* also contains a part that is called the *Weyl curvature tensor*. Even without the presence of sources the Weyl curvature tensor represents that part of space–time curvature which can curve up a void. A vacuum space–time is therefore not necessarily flat.

Looking at this space–time from the point of view of an evolving empty space, we speak of a three-dimensional hypersurface (the space) embedded into this four-dimensional space–time tube. Geometrically, we have just split the space–time into space + time, so that we can talk about an *evolution of space*. This is the principle structure of a cosmological model, even if there is no matter.

It turns out that this space is characterized by a negative scalar curvature, while it is expanding (or contracting), (see also the discussion in the appendix). This is a consequence of splitting the space–time (Ricci) curvature of the tube (which vanishes for an empty region) into a three-dimensional so-called *intrinsic curvature* of space and a so-called *extrinsic curvature* of the embedding of the space into the space–time. One may say that the former describes the space as seen from "inside", and the latter as seen from "outside" within the four-dimensional space–time. This latter curvature can be formally interpreted as an expansion (or contraction) of space (together with other kinematical properties). We could also say that this curvature "appears" as expansion to an observer; physically it is just a geometrical curvature describing how the space is embedded into the space–time. If thought through properly, this geometrical picture clarifies what we could mean by an expanding Universe, since this picture also applies to a model with matter. In the case of no matter the clue is that both curvatures (the intrinsic and extrinsic) have to compensate each other in order to give a vanishing (Ricci) curvature for the space–time tube. In other words, we created an expanding (or contracting) space with a negative curvature out of a vacuum.

This example illustrates a general feature of the cosmological point of view: in cosmology one thinks of a space that evolves in time, while the theory of general relativity describes four-dimensional space–times. The above-described "embedding problem" is a philosophically interesting issue, since we have some freedom of what we call space: we may cut the space–time differently. This touches on a deeper problem of what we wish to consider as "space" or as "time". If we just look at a three-dimensional space geometrically, we may observe its evolution and parametrize the different evolution steps by some parameter t, called the time, and then stack all these spaces together into a space–time tube, that contains this sequence of the evolving spaces (Fig. 1.17). If we do this with some generality it turns out that the "time" is *local*, that is, only in special cases we would have a global universal time for each point in space. These local times would depend on the structure at these space points too. This example shows that Einstein's theory can have rich consequences and the details are much more involved than the simple standard model would suggest.

Fig. 1.17 We can create a space–time tube by stacking together evolving spaces. We look at a spatial three-dimensional hypersurface (here drawn as a two-dimensional surface) and label the different moments in its evolution by a parameter t, and then draw this evolving sequence of spaces within a space–time tube. In this picture the parameter t can be interpreted as a universal "time". There are, however, more general cases, where the time is only local and depends on the structure at all points in space [Credit: Mauro Carfora]

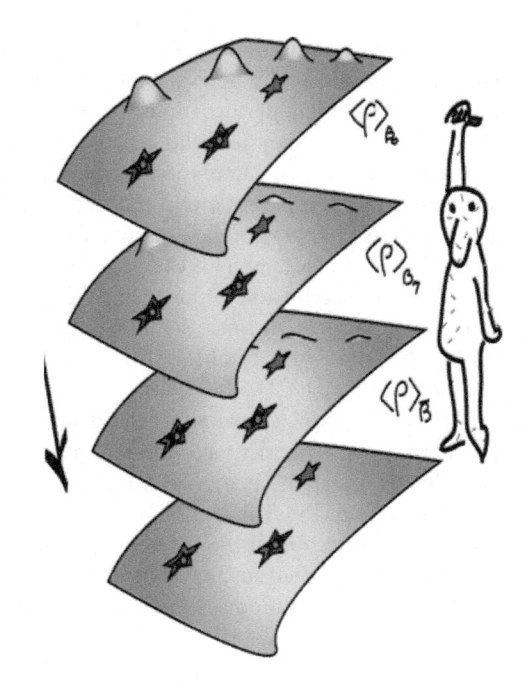

1.5.3 Difficulties of the New Paradigm

If we understand that a more realistic model would mimic an effect that acts in the right way to explain the dark mysteries of the old universe model, why are many researchers not convinced yet? There are several possible answers: the standard model has a long history and it was very robust in the past surviving the many new observational results that we had: it is still today an excellent model within which many of these observations match together – well, apart from some yet unexplained anomalies and, of course, were there not these dark components needed! One has to be very careful with this statement. The standard model is often assumed *a priori* and, before people look at observations, they often assume the standard model to hold. Changing a model needs a large effort of reinterpreting all these observational data. In the new model, almost everything changes and standard methods like simulations of structure, built on Newton's theory, cannot be used any more. This amounts to nothing less than a paradigmatic change of cosmology. Geometry, the basis of Einstein's theory, has to be taken seriously and a lot of new methods have to be developed to reach the high standards of interpretation within the standard paradigm.

Another answer has to do with a big open question: although we know a correct qualitative answer, it is not clear whether all this gives an effect large enough to match the needs for Dark Energy and Dark Matter. People are currently working hard to resolve this, but most of them are working as close as possible to the standard paradigm: the new mechanism and the role of curvature is explored, but starting with small perturbations of the old flat standard model. Scientists may expect too much: they find the effect to be insufficient, but this may be the result of their conservative calculations, assuming that the evolution of the Universe is very close to the standard model. Moreover, as David Wiltshire from the University of Canterbury in New Zealand pointed out recently, the observer looks at structures from a special place: he is hosted in a galaxy and this place does not even partake in the general expansion of the cosmological model. Additional effects, also related to the observer's *local time*, that developed differently in different regions of space during the whole history of the structure formation process, would have to be taken into account (see [22]).

Another group of researchers considers special exact solutions of Einstein's equations (with spherical symmetry), roughly describing a situation that we accidentally live near the center of a huge void, an idea that has been spelled out by Kenji Tomita of the Yukawa Institute in Kyoto, Japan. These special solutions are not able to describe all the effects we discussed, but people succeeded already to explain observational results without Dark Energy in these models. Even if these models and the assumptions made may appear special, these are very encouraging results.

At present, all these model considerations are not conclusive. The research field is too young to expect this. All the points mentioned must be carefully investigated. Or, perhaps, a more dramatic change has to happen, and we now turn to a view that would imply such a more dramatic change concerning the global structure of our Universe.

1.6 Construction Principles for a Cosmological Model

Before entering the endeavor of constructing a cosmological model, it is good to think of *principles* that should lie at the basis of this construction. They should be formulated such that we are able to falsify them by confronting the constructed model with observations. Such principles are actually needed for a very simple reason: we are in principle not able to see the whole Universe, if its size exceeds what we call the *horizon*. We can observe only a fraction of our Universe, since we draw information from the light emission of objects, and light can only travel at a limited speed. We already said that the light follows the space–time geometry. Assuming that the Universe had a beginning, then – due to the limited travel speed of information – only a finite volume of the Universe can be seen: there is a horizon, beyond which we cannot obtain any information of our Universe at present, and if we wait a little, then we would have a little more information, but never all. The standard model can actually be derived from a principle that circumvents this problem and that we explore now.

1.6.1 A Strong and a Weak Version of the Cosmological Principle

According to this principle, called the *strong cosmological principle*, the problem mentioned above can be cured by assuming that all observers in the Universe would see the same in all directions as we do, and so the postulate was born that the properties of the standard model (that does not distinguish a preferred place in the Universe) would hold everywhere. In concrete terms, all what we see up to the horizon is what another inhabitant of the Universe would see up to his/her horizon. This suggests to extrapolate our regional properties to all of the Universe: if we see an expansion, then space is expanding also beyond our horizon and, according to the standard model, the Universe will never and nowhere cease to expand. Mathematically, this cosmological principle can be formulated in a strong way: if we assume that, at every point, the universe model is isotropic, that is, it does not prefer any direction, then the model must also be homogeneous. This principle, thus, establishes the standard homogeneous-isotropic model.

The currently held cosmological principle appears, from a physical point of view, very strong: the matter distribution as we see it is far from homogeneous and also far from isotropic on smaller scales. On large scales it may obey these properties in some averaged sense. All this points to the need to relax the *strong cosmological principle* and replace it by a weaker version. The *weak cosmological principle* assumes that, on very large scales of billions of lightyears, the Universe is isotropic and homogeneous, if the distribution of matter and geometry is averaged over. Beyond this scale nothing new happens, according to this principle, and we may explore this large-scale portion of the Universe only and extrapolate the inferred properties to the whole Universe. Note that even this weak form of the cosmological principle demands something very strong: it may be that the structure in the Universe is such that there exists no scale beyond which its properties would not change. This possibility has been reflected and analyzed by Francesco Sylos Labini of the University of Rome in Italy. Nevertheless, this is a first step to generalize the cosmological principle. The standard model obeys the strong version, the generalized cosmological models could be classified into two: those that do satisfy the weak principle and those that do not. Consequently, we shall first consider those models that are on large-scales isotropic and homogeneous, which means that the observational constraints that we have on isotropy and homogeneity are also met by the more general models.

Interestingly, models that would obey the *weak cosmological principle* are rather conservative. People are currently working mostly on this class of models. However, it is illustrative for our understanding to consider a model that does not obey this principle, although it is very simple. This cosmology is a generalization of Einstein's static cosmos by including inhomogeneities in the Universe.

1.6.2 What Einstein Wanted

Let us remember Einstein's static universe model. It was abandoned clearly on the grounds of an observed recession of the galaxies away from us together with the explanation in terms of an expansion of space (see, however, [12]). But what would happen if our Universe were really static, but in a much larger volume than that accessible by our observations? Could the expansion of space be a regional property that changes at other places? Exactly this was brought into the discussions around Einstein's model by Sir Arthur Eddington at the time where people started to prefer Lemaître's expanding universe models. He said, [10], ... *that it is possible that the recession of the spirals is not the expansion theoretically predicted; it might be some local peculiarity masking a much smaller genuine expansion; but the temptation to identify the observed and the predicted expansions is very strong.* Of course, a homogeneous, static solution does not describe expansion also in smaller regions, and therefore it is clearly not an appropriate model. However, as we explained, a more realistic model has fluctuations in matter and geometry and we may ask Einstein's question afresh: could a generalized averaged model be static on a very large scale and at the same time predict an expanding space in some region around us? Not surprisingly, the answer is yes. Let us look again at our (1.2), but this time we apply it to the whole Universe:

$$\text{global acceleration} = - \text{ density of all matter} + \Lambda + \text{global } Q. \quad (1.3)$$

In Einstein's model, the global $Q = 0$ (it is homogeneous) and the total acceleration of the volume expansion is also vanishing together with its expansion. We have the condition that the cosmological constant (introduced by Einstein to realize such a cosmos) must balance the whole matter content of the Universe:

$$\text{density of all matter} = \Lambda. \quad (1.4)$$

We learn from this equation Einstein's reason for introducing the cosmological constant: without it the density has to vanish, so that a realistic Universe with matter cannot be static.

Now, in the new model we can also construct a Universe with no acceleration on the largest scales. Then, even without a cosmological constant, we can balance the total matter content by their own fluctuations Q:

$$\text{density of all matter} = \text{global } Q. \quad (1.5)$$

We could see this condition as a sort of equilibrium around which the Universe is hovering: matter is not distributed in a calm and homogeneous way, there are structures and motions of these structures, but altogether the available energy in matter and the energy in its motions balance each other. In this cosmos, which is overall static, we nevertheless find deviations from a homogeneous distribution, so that there are smaller regions in the interior of this cosmos with expansion, alternating with regions that are in the state of contraction.

1.6.3 A Universe That Looks Like a "Multiverse"

The above universe model realizes in a sense what Einstein wanted to have: a globally static Universe within which things evolve. His model was just too special to describe an evolving interior, since it was homogeneous, i.e., at all places the space was static. Now, in the more general, globally static model the interior is changing all the time with a pulsating alternation of over- and under-dense regions. Since these regions could be large and of the order of the observable Universe (the horizon), we could imagine this model like a (connected) "Multiverse": our Universe corresponds to a region, that is, in the state of expansion, and this state must even accelerate according to this model. We may draw an analogy with the geoid of the Earth in Fig. 1.18 to strengthen our intuition.

Imagine that we are living in such a regionally expanding and accelerating part of the Universe. When looking back in time we would see a state of non-accelerated expansion and, looking further back, a state where the expansion emerged from a quasi-static state. Nearby, i.e., on the scale where we have obtained the galaxy catalogues, we would see a rather uniform expansion, but further out it will be different than in the standard model, where the expansion speed is the larger the further we look, here the expansion must slow down at some point. Note that it is very difficult to justify such a non-uniform expansion from observations, since we observe redshifts and not distances. These latter are only obtained by having a model, so if the model is different, then the interpretation of the redshift is different. If the region of expansion would be smaller we could expect to see an overabundance of galaxies, a "blue excess", since the number of galaxies per redshift interval would increase more than expected from a uniform distribution in the standard model.

This example of a very different cosmology shows that it can be interesting to relax the cosmological principle. The model discussed here is difficult to justify as a valuable model for all the observations we have – simply also because one has not attempted to compare it with all these observations – but it would have no need for Dark Energy: in an under-dense, expanding space region we find that Q is larger than the density of matter, which is exactly the property needed for acceleration of the space volume. (A discussion of the equations corresponding to this model can be found in the appendix.)

1.6.4 How Einstein Argued That the World Cannot Be Flat?

This model is, however, far from the standard model of cosmology. It is not a perturbation of the standard model. It implies that our world is not flat or almost so (i.e., quasi-Euclidean), but that it has a dominant global curvature: the world as a whole is positively curved, like the Earth, the Sun and all things we know to exist in this world, and the favorite space-form of Einstein and Eddington is

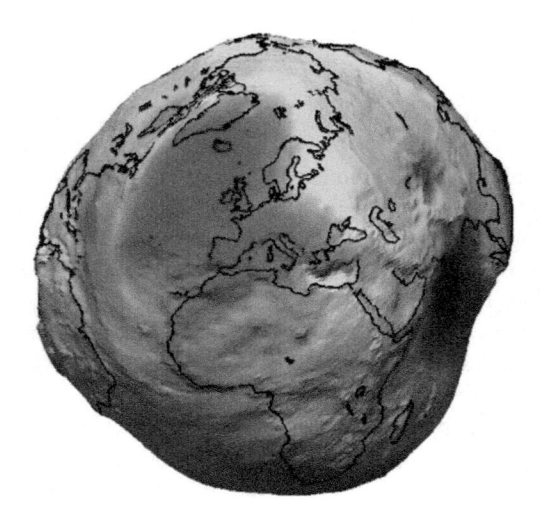

Fig. 1.18 This picture draws an analogy of the global structure of a spherical Universe by a two-dimensional analog of the Earth's surface. The Earth is commonly idealized as an exactly spherical ball, while in reality there are structures that would cause deviations from this ideal geometry. Looking at the gravitational field at the surface of the Earth, geologists draw the so-called *geoid*. It defines a surface on which, at any point, the pull of gravity is perpendicular to it. If we would put a ball on this hypothetical surface, it would not roll, even though the surface appears to have slopes. This is because the geoid surface is constructed such that the gravitational field is exactly equal at every point on the surface. This modeled geoid is highly exaggerated to illustrate the point. In the real Universe, that could have a three-dimensional spherical space form, the irregularities in space (corresponding to the irregularities shown by the geoid in two dimensions) are actually very tiny: the curvature may be relatively weaker in some regions of space corresponding to relatively empty regions, and it may be relatively stronger at places where the regions are filled with relatively more clusters of galaxies. Overall, the model would be static, but regionally there are expanding and contracting regions. These regions may be as large as the horizon. Estimates show [4] that the universe model discussed here would contain at least 50 horizon volumes. That leaves enough room for an alternation of expanding and contracting regions. That we draw the analogy with the geoid is due to the fact that the geometry of space in Einstein's theory traces the gravitational field generated by the structures. Notice that this model does not obey the weak cosmological principle, since we cannot extrapolate the properties of a region, for example an expanding region, to the whole Universe [Credit: GFZ Potsdam; http://icgem.gfz-potsdam.de/ICGEM/]

recovered. Let us look at a remark by Albert Einstein himself who argues for a curved, finite space form already in 1916, after he recalls what we now name the standard model of cosmology [11]: *...But it is conceivable that our Universe differs only slightly from a Euclidean one, and this notion seems all the more probable, since calculations show that the metric of surrounding space is influenced only to an exceedingly small extent by masses even of the magnitude of our Sun. We might imagine that, as regards geometry, our Universe behaves analogously to a surface, which is irregularly curved in its individual parts, but which nowhere departs appreciably from a plane: something like the rippled surface of a lake. Such a Universe might fittingly be called a quasi-Euclidean Universe. As regards its space* *it would be infinite. But calculation shows that in a quasi-Euclidean Universe the average density of matter would necessarily be zero. Thus, such a Universe could not be inhabited by matter everywhere. [...] If we are to have in the Universe an average density of matter, which differs from zero, however small may be that difference, then the Universe cannot be quasi-Euclidean. On the contrary, the results of calculation indicate that, if matter be distributed uniformly, the Universe would necessarily be spherical [...]. Since in reality the detailed distribution of matter is not uniform, the real Universe will deviate in individual parts from the spherical, i.e., the Universe will be quasi-spherical. But it will be necessarily finite.* Later, in 1952, Einstein added the following remarks [11]: *I further want to remark that the theory of expanding space, together with the empirical data of astronomy, permit no decision to be reached about the finite or infinite character of (three-dimensional) space, while the original 'static' hypothesis of space yielded the closure (finiteness) of space.* It should be added that, nowadays, people have resolved the doubts raised by Einstein: one constructs a quasi-Newtonian model on top of the homogeneous-isotropic standard model by introducing deviations in the density that could average out to zero within a periodic box. We have explained this architecture in Sect. 1.3. Although technically this construction can be shown to be correct [5], it nevertheless must be considered artificial: the Universe is conceived as a replication of identical boxes and the structures are imprisoned into those boxes having a vanishing average, so that at the end the boxes (and the whole Universe) would behave, on average, like the standard model. One could say, this is more by construction rather than by nature, and it is certainly worth listening to Einstein and rethinking this construction.

The model presented above agrees in spirit with what Einstein imagined for a realistic universe model. It would be more difficult to argue why this model should be abandoned. At least, this is an example that there is a much larger spectrum of possible cosmological models in the framework of Einstein's general theory of relativity than the old family of homogeneous models of Friedmann and Lemaître. Another example may illustrate this: the balance condition used to construct a fluctuating Einstein cosmos does not necessarily imply that the model is static. We may even construct, with the same principle, a model that is globally expanding. Still, the standard cosmological principle is violated: in such an expanding stationary cosmos there are accelerating regions and decelerating ones, and we could say that we happen to live in an accelerating region. Again, there is no need for Dark Energy, and by measuring the Hubble expansion in our region we could probably not even distinguish this model from a homogeneously expanding standard cosmology with a cosmological constant. (More details may be found in [4].)

1.6.5 A Matter of Principle

These examples show that it is a matter of principle on how we look at the Universe: the above models impose a global criterion, the balance of the matter content and its fluctuations. This balance could be a sort of equilibrium: researchers who study biological systems like a living cell call such a balance a far-from-equilibrium state. It is characterized by a constant in- and outflow of matter, while maintaining a situation in which things could keep their identity. In such a system information in the form of differentiation of structure is produced. On the contrary, if we suppress these in- and outflows, the cell is closed and it consequently dies. This latter is called thermodynamical equilibrium. All this is in large contrast to the standard model which would in this picture correspond to a dead cell. The standard model is based on a local criterion of the way how the Universe should look around us. But, as we know now, such a principle is too strong: the Universe around us actually looks very different than the standard model suggests.

Cosmology still is a philosophical science, since we have to impose criteria or physical principles in order to construct our models and then confront them with observations. We have to be openminded and admit that a model other than the standard one may explain our observations. Many cosmologists argue that there is no need to change the standard model, because it works very well. But, we must reply: why does it need 95% of Dark Energy and Dark Matter? If the physical origin of these dark sources is not found, the standard model simply does not work. Conceiving such a more general universe model, it is also more difficult to determine it. While the standard model just needs three numbers to determine its evolution (the density of matter, the homogeneous curvature at some instant of time, and the cosmological constant), the more general models are much more complex and they do evolve much more freely.

Related to the task of finding principles to build cosmological models is an interesting discipline of mathematics, that may be key also for the construction of cosmological models. We therefore dedicate the final section to the research field of *cosmic topology*.

1.7 The Global Shape of the Universe

Developing mathematical models based on physical laws, we are used to think about fields that describe physical properties like the electromagnetic field, but also geometrical properties, like the gravitational field in Einstein's theory. The laws studied are expressed in terms of differential equations. Also the geometry is expressed in terms of differential geometry. There are, however, other possible structures that are not formulated in a differential way, notably structures that describe the connectivity of space. Imagine that we have a surface that is described geometrically, and suppose we have a structure on this surface that allows us to measure lengths of a path along which we may walk. Imagine now that we say that two points on this path are identical, which imposes a property of connectivity. As soon as we hit one of these points, we would be able to switch to the other point on the path. We would for example move along the path to the first point and "re-enter" the surface at the other point. All this brings us to the mathematical discipline of *topology*.

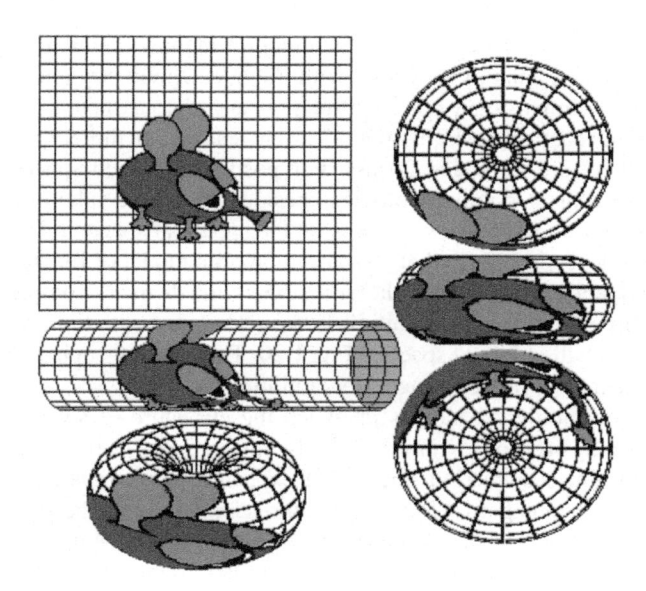

Fig. 1.19 The two-dimensional torus can be constructed as an embedding into the three-dimensional space by folding and identifying opposing faces. The geometry of the surface, however, remains flat. Analogously, the three-dimensional hypertorus could be constructed and embedded into the four-dimensional space. The torus topology is an additional structure of connectedness that identifies points in the geometry [Credit: Karl Bednarik; GNU Free Doc. Lic.; http://de.wikipedia.org/wiki/Datei:MUFFIL-1_Verheftung_und_Form_des_2-Torus.png]

1.7.1 Topology as a Key-Discipline of Cosmology

The topology of space may be the key to approach the construction of a universe model. We already mentioned two examples: the universe model favored by Einstein and Eddington is a spherical space with positive curvature. Such a space form has spherical topology and a finite volume. Also, the architecture of current structure formation models, discussed in Sect. 1.3, sets out a specific topology, since periodic replica of boxes are nothing else than a single box, where all points on the opposing faces are identified.

Imagine that you are moving in the box and finally hitting the boundary. If the structure inside the box is periodic you will either leave the box to see the same structure in the neighboring box (topologists call this point of view the *covering space*), or you may just stay within the same box and re-enter it from the opposite side (this is called the *fundamental cell*). The identification of points in space does not change its geometry, but its connectivity. You can build a simple model of this so-called torus topology: you take a square for simplicity that is made up of a deformable material. Then you coil one boundary of this two-dimensional box onto the opposing boundary: you first get a cylindrically shaped form. Then, you do the same with the remaining two boundaries and you obtain a donut-shaped form, see Fig. 1.19.

On this donut you can move around without hitting any boundary. Accordingly, in three dimensions, a flat hypertorus – with Euclidean geometry but with opposing faces of a three-dimensional box identified – can be constructed. In a flat space we can introduce other rules of identification leading to finitely many possible topologies (there are 18 space forms). For positively curved spaces there is an infinite number of possible topologies: besides the three-sphere you can, for instance, identify all the opposing points on the spherical space and you obtain the so-called elliptical space form that was also a favorite model of Einstein.

For negatively curved spaces the situation is much more complex: there are infinitely many space forms, with infinite volume, but also with finite volume as, for example, the so-called Picard space of Fig. 1.20. This space form has finite volume, but it features one "horn" that reaches out to infinity.

There are cosmological models in which the curvature of space may change: the Universe may start out with a positively curved space-form similar to that of the static Einstein cosmos, but due to the expansion the curvature of space will become flat and then negative. Such a curvature change can be the signal of a much more dramatic change of the shape of the Universe: its topology, describing how different regions of space are connected, can also change: a negative curvature Universe with a hyperbolic (saddle-shaped) topology must have disconnected regions and, if its volume is finite, a generic space-form exhibits so-called "horns" that reach out to infinity. Its shape is that of a "hedgehog" (see Fig. 1.20 for a special example). Again, this is reminiscent of a biological cell which also undergoes topological changes in the form of cell divisions.

1.7.2 Can We, in Principle, Determine the Topology of Our Universe?

Imagine that we pick a certain spatial scale in our Universe, say a region that contains as much as a cluster of galaxies. Then, proceed by choosing a topological space form that is compatible with the geometry of that region, and select one of the possible space forms. You so claim that the Universe is like a kaleidoscope, a repetition of this single cluster of galaxies. You create mirror images of the galaxies in the covering space. You can then say that a far away galaxy hosted within a – just apparently other – cluster of galaxies is in fact one and the same galaxy in the same cluster of galaxies. With this picture in mind you may then think of strategies of how you could observe the fact that you see one and the same galaxy. It is clear that this would be in principle possible, since the light could have travelled several times through the fundamental cell of this cluster of galaxies, so that you would observe the same galaxy several times from different angles, but also at different evolutionary stages. Both of these latter facts ask for quite sophisticated strategies as you can imagine. The resulting picture also depends on your choice

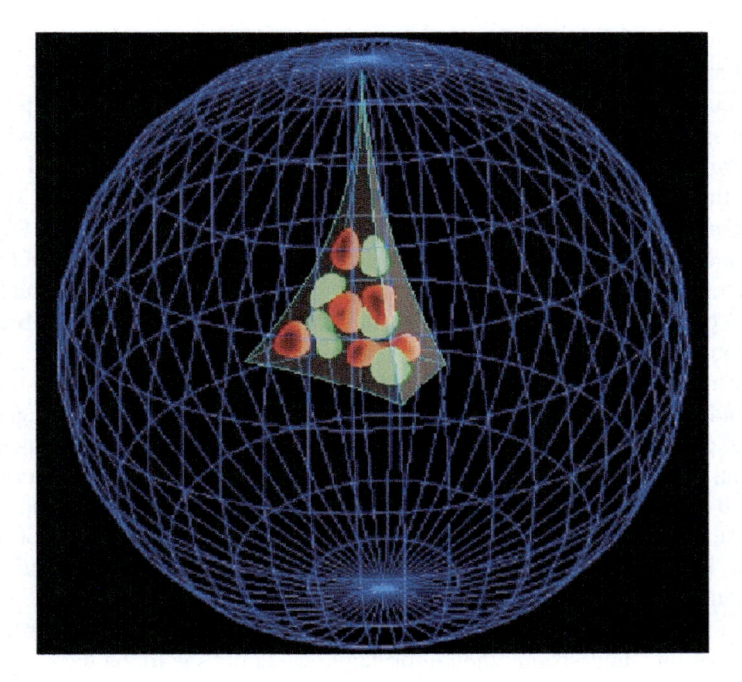

1 Dark Energy and Dark Matter Hidden in the Geometry of Space? 35

Fig. 1.20 A possible space-form with hyperbolic geometry and finite volume is the so-called Picard space. It has just one "horn" reaching out to infinity, but having a finite volume. Topologists call the pyramid inside the unit ball (in yellow) a "fundamental cell" that generates the topology of space. The colored blobs within this cell indicate possible oscillations: it is like a drum that allows only for special oscillations (sounds), and this is because the drum is finite like this universe model. Frank Steiner of the University of Ulm and his collegues have shown that our observations of the Cosmic Microwave Background are in better accord with this model, since observations indicate that oscillations on a scale larger than a certain finite volume scale seem to be missing, which is naturally explained here. In general we have to be aware that many such horns could be created: the Universe looks like a "hedgehog". Such a space form does not obey the "cosmological principle" that the Universe looks everywhere the same: one inhabitant could even live up the "horn", while others would reside in a region of an accelerating space, while again others could see a decelerating space around them. Those that see an accelerating domain would be us, and if we would use the standard model to describe our region, we would need Dark Energy to explain this observation. Here we do not [Credit: Reprinted Fig. 2 from [1] http://prl.aps.org/abstract/PRL/v94/i2/e021301 ©(2005) American Physical Society]

of what you think is the fundamental cell. If the fundamental cell is as big as the observable Universe, (the horizon that we introduced earlier), then we are led to consider another important possibility.

1.7.3 Topology of the Cosmic Microwave Background

This possibility, that has been followed to some depth in the cosmological community, is to look closer at the Cosmic Microwave Background maps. The Universe saw a time in the past where the cosmic substratum of matter was optically thick, i.e., radiation could not escape. It is at the time when atoms recombine, so that the free electrons can no longer scatter the radiation, and so light can freely propagate through space. It is that moment, or better a certain range of moments, where the light forms a thick shell, being spherical in the isotropic Universe, and centered around the observer that should have no preferred position in space, a so-called *fundamental observer*. We are not such fundamental observers, since the Earth moves around, together with the Sun that moves around the galactic center and this center moves towards the local cluster of galaxies. But we can construct a map centered on a fundamental observer by removing that part of the Cosmic Microwave Background radiation that is caused by our proper motion.

This motion creates a Doppler effect, similar to the experience of listening to the sound of a car that passes by: if the car approaches the sound is higher, and if it has passed the sound becomes lower in frequency. The same happens with radiation, so that we are seeing in fact a hotter region before us and a colder behind. Such a dipole we can remove from the map upon claiming that this dipole seen in the microwave sky is all due to our proper motion (that may or may not be fully the case, however). The remaining signal is a quadrupole and higher multipoles of the radiation that provides us with a map of the celestial microwave sphere. The clue is now: if the Universe is finite and has a diameter that is smaller or of the order of the diameter of this light-sphere, then in principle we could see topological signatures of our Universe. One such signature is due to the finiteness of space, as illustrated in Fig. 1.21. Analyzing the auto-correlation of the patterns seen in the microwave maps would then reveal a suppression of correlations as soon as we approach the scale of the Universe.

1.7.4 The Search for "Matching Circles"

It is interesting that the analysis of many different finite universe models, if compared to current data of the Cosmic Microwave Background, always points to a typical size of about 4 Hubble-lengths which, if raised to the cube, would be compatible with what we estimated for the volume of the globally static universe model, *cf.* Fig. 1.18. We would of course want to know the topology more precisely than just the property of being finite or infinite. Here, cosmologists have also developed strategies that search for "matching circles" in the sky [9], see Fig. 1.22. In practice, however, such a strategy is very difficult, since we do not know precisely which topological space form to choose (and there are infinitely many), and we do not know precisely the size of the fundamental cell. However, people are smart and they will eventually conduct comprehensive programs to finally reach a realistic possibility to at least determine classes of topological space forms that are excluded by the data.

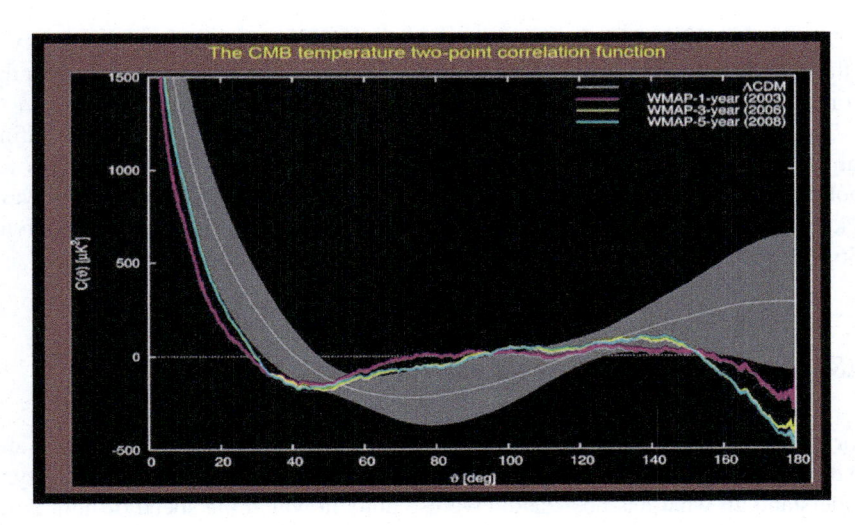

1 Dark Energy and Dark Matter Hidden in the Geometry of Space?

Fig. 1.21 The auto-correlation function is a simple statistical method to compare observational data of the Cosmic Microwave Background with cosmological models. The white line shows the standard concordance model with a gray area that takes into account the possible variations around the mean expected value of the model. The colored lines show the observational data for three successive measurements of the satellite WMAP (Wilkinson Microwave Anisotropy Probe). We see that the correlations on large scales are significantly suppressed in the observations. We may interpret this in terms of finiteness of space: a model with a finite space simply does not have structure on scales beyond the finite volume. It is then in better agreement with the data (see: [2]) [Credit: Frank Steiner]

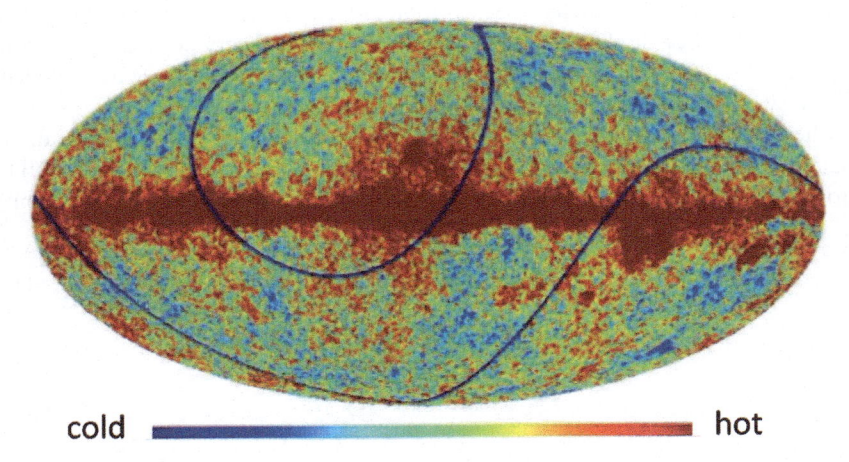

Fig. 1.22 The search for "matching circles" in the cosmic microwave sky furnishes a possibility to determine the topology of our Universe. If the diameter of the Universe is smaller than the diameter of the actual light-sphere, then replica of the space (in the spirit of the simple torus universe model) would each contain a light-sphere. Two neighboring light-spheres would then overlap along circles, on which the pattern seen in the maps should be identical. (The dark red bar in the equatorial plane shows the emission of our galaxy that has to be masked for a proper analysis of the primordial radiation.); see [9, 26] [Credit: Frank Steiner]

The Universe in which we live may hide some more surprises, and cosmology is a thriving discipline of science that certainly does not end with the picture of a flat world containing mysterious energies.

The reader may find some discussions in popular articles about this exciting subject, for example in [23–25] and [26], and some interesting internet sites for looking at galaxies and galaxy catalogues [27–32], maps of the Cosmic Microwave Background [33], topology [34], numerical simulations [35], general relativity [36, 37], and experimental and theoretical issues on warp drive [38, 39].

1.8 Appendix: The Equations Behind the Words

For the reader who is more familiar with equations, the following sections are useful to deepen what has been said in the text. This material is, however, not necessary to understand all what has been said. (We use units in which the speed of light is set to unity.)

1.8.1 The Standard Model of Cosmology

Alexander Friedmann has first derived a simple class of solutions for Einstein's equations. His equations impose a strong restriction on Einstein's equations, by assuming matter and geometry to be *isotropically distributed about every point*, i.e., his universe models look the same in all directions from every point in space. Such an assumption can only be realized, if the matter distribution in space as well as the geometry of space itself are both completely homogeneous.

The resulting equations can be written as an *expansion law*, that is the temporal change of a scale-factor $a(t)$ of the Universe (which is a function of time like all the variables), and an *acceleration law*, that is the temporal change of the expansion:

$$\left(\frac{\dot{a}}{a}\right)^2 = \frac{8\pi G \rho_h}{3} + \frac{\Lambda}{3} - \frac{k}{a^2}; \qquad (1.6)$$

$$\left(\frac{\ddot{a}}{a}\right) = -\frac{4\pi G(\rho_h + 3p_h)}{3} + \frac{\Lambda}{3}; \qquad (1.7)$$

$$\dot{\rho}_h + 3\left(\frac{\dot{a}}{a}\right)(\rho_h + p_h) = 0. \qquad (1.8)$$

In the first equation, $\frac{\dot{a}}{a} = H(t)$ is the relative rate of change of the scale factor $a(t)$, i.e., the expansion, that sometimes is denoted by the Hubble function $H(t)$, ρ_h is the density of matter, i.e., the mass per unit volume, and p_h the pressure (where the index h stands for homogeneous), the term Λ is the cosmological constant introduced by Einstein into his equations, G is the universal gravitational constant, and k is a constant that describes an eventual homogeneous curvature of space, i.e., a curvature that is everywhere the same (like in the case of a sphere in two-dimensions). This constant can be zero (a flat space), positive (a closed spherical space) and negative (a hyperbolic space). In the second equation we can observe that the cosmological constant, if it is positive, can accelerate the expansion of the Universe and so can counteract gravitation. The matter density is always positive, so it always slows down the expansion (it has a negative sign in front of the acceleration equation). The third equation, finally, connects the two others and describes the conservation of energy during the evolution. The three equations are connected in the sense that, if the third equation holds, then the second equation is just the time-derivative of the first.

The parameters of the model are usually written by dividing the expansion law above (the first equation of the set (1.6)) by the square of the Hubble function H^2. Then, one obtains a sum of three cosmological parameters:

$$\Omega_m + \Omega_k + \Omega_\Lambda = 1, \qquad (1.9)$$

with the definitions:

$$\Omega_m := \frac{8\pi G \rho_H}{3H^2}; \quad \Omega_k := \frac{-k}{a^2 H^2}; \quad \Omega_\Lambda := \frac{\Lambda}{3H^2}. \qquad (1.10)$$

These three parameters can be determined from observations.

In the *Concordance Model* the Universe should have the following properties:

(a) It is spatially almost flat, that is $k \approx 0$, and the parameter Ω_k is very small and usually set to zero so that it disappears from the above equations.
(b) Known matter and Dark Matter together would make up a matter parameter of $\Omega_m \approx 0.27$, where only a part of about 0.04 is made of known (so-called baryonic) matter and only a small fraction of it, is shining baryonic matter (radiation itself and neutrinos are here considered as negligible).
(c) The cosmological constant Λ is positive and employed as the simplest model of Dark Energy [19]. The corresponding parameter is then $\Omega_\Lambda \approx 0.73$ and must fill the remaining contents of the Universe, since the sum of all the parameters must be equal to 1 to satisfy Einstein's equations.

1.8.2 Averaged Cosmological Equations

In order to find corresponding evolution equations for spatially averaged cosmological variables, we may put the following simple idea into practice. We observe that Friedmann's expansion and acceleration laws rely on the strong symmetry assumption of local isotropy that implies a homogeneous model. Hence, by dropping the strong symmetry requirement, we obtain more general equations. In particular, we can take the general Einstein equations for the expansion and the acceleration of the volume of the Universe and average them over, i.e., we integrate a scalar variable over a spatial domain $\mathcal{D}$, and divide this integral by the volume of the domain $\mathcal{D}$. We so obtain very similar equations but with extra terms that reflect the inhomogeneity and anisotropy of a general distribution of matter and geometry. We here give these equations, for the sake of simplicity, for the case of vanishing pressure only [3]:

$$\left(\frac{\dot{a}_\mathcal{D}}{a_\mathcal{D}}\right)^2 = \frac{8\pi G \langle\rho\rangle_\mathcal{D}}{3} + \frac{\Lambda}{3} - \frac{\langle R\rangle_\mathcal{D} + Q_\mathcal{D}}{6}; \qquad (1.11)$$

$$\left(\frac{\ddot{a}_\mathcal{D}}{a_\mathcal{D}}\right) = -\frac{4\pi G \langle\rho\rangle_\mathcal{D}}{3} + \frac{\Lambda}{3} + \frac{Q_\mathcal{D}}{3}; \qquad (1.12)$$

$$\langle\rho\rangle_\mathcal{D}^{\cdot} + 3\left(\frac{\dot{a}_\mathcal{D}}{a_\mathcal{D}}\right)\langle\rho\rangle_\mathcal{D} = 0; \qquad (1.13)$$

$$\frac{1}{a_\mathcal{D}^6}(Q_\mathcal{D} a_\mathcal{D}^6)^{\cdot} + \frac{1}{a_\mathcal{D}^2}(\langle R\rangle_\mathcal{D} a_\mathcal{D}^2)^{\cdot} = 0. \qquad (1.14)$$

We have replaced the scale-factor of the homogeneous Universe by the *volume scale factor* $a_\mathcal{D}$, defined through the ratio of the volume of a chosen spatial domain $\mathcal{D}$ and its volume at some initial time:

$$a_\mathcal{D}(t) := \left(\frac{V_\mathcal{D}(t)}{V_{\mathcal{D}_i}}\right)^{1/3}. \qquad (1.15)$$

This scale-factor effectively describes the volume expansion, but since the volume is not the volume of a spherical region as in the standard model, but is deformed due to the inhomogeneities, the scale-factor is in principle different in different directions. Taking the volume is therefore simpler and also correct. An important difference, however, is the occurrence of the index $\mathcal{D}$ that denotes the domain in space over which one has averaged the distribution. If we choose another domain $\mathcal{D}$, then the volume has a different evolution. We see here that the new model is already more general in this respect, since it takes into account the fact that the evolution on different spatial scales must in general be different. In the standard model, every domain evolves in the same way.

We can proceed as in the standard model and construct cosmological parameters by dividing the general expansion law by $H_\mathcal{D}^2$, where $H_\mathcal{D} := \dot{a}_\mathcal{D}/a_\mathcal{D}$, which reduces to Hubble's function in the homogeneous case. We so obtain four parameters that have to sum up to 1:

$$\Omega_m^\mathcal{D} + \Omega_\Lambda^\mathcal{D} + \Omega_R^\mathcal{D} + \Omega_Q^\mathcal{D} = 1, \qquad (1.16)$$

with the definitions:

$$\Omega_m^\mathcal{D} := \frac{8\pi G \langle\rho\rangle_\mathcal{D}}{3H_\mathcal{D}^2}; \quad \Omega_\Lambda^\mathcal{D} := \frac{\Lambda}{3H_\mathcal{D}^2}; \quad \Omega_R^\mathcal{D} := -\frac{\langle R\rangle_\mathcal{D}}{6H_\mathcal{D}^2}; \quad \Omega_Q^\mathcal{D} := -\frac{Q_\mathcal{D}}{6H_\mathcal{D}^2}.$$

$$(1.17)$$

If we compare the corresponding equations of the standard model with these new equations, we see that there are new terms appearing. These so-called "backreaction terms" we explain now, together with other important differences.

1.8.3 Discussion of the New "Backreaction Terms"

Looking at the new cosmological equations we can spot two new terms. The first term, $Q_\mathcal{D}$, is called the *kinematical backreaction* that arises as a result of the inhomogeneities. The term *backreaction* means that these terms would react back on the evolution of a model that is homogeneous. In detail, this extra term is written:

$$Q_\mathcal{D} := \frac{2}{3}\left\langle(\theta - \langle\theta\rangle_\mathcal{D})^2\right\rangle_\mathcal{D} - 2\langle\sigma^2\rangle_\mathcal{D}. \qquad (1.18)$$

The first term in this expression describes the fact that the expansion, called θ at every point in space is different from the average over the expansion $\langle\theta\rangle_\mathcal{D}$ at all points in $\mathcal{D}$. The expansion at every point would be equal to the averaged expansion only in the homogeneous standard model, and therefore this term is not there in the equations of Friedmann. The second term in the above expression is the so-called shear that measures the anisotropy of the structures. Also this term vanishes in the standard model, since isotropy is required, i.e., vanishing shear. On large scales this term could be subdominant if compared to the first expansion term, since at some large scale the structures are no longer anisotropically distributed and, although the average sums up all the contributions up to the scale of averaging, the average is divided by the volume that will hence suppress the shear contributions present on smaller scales. In concrete models it turns out that this is indeed the case and the overall term $Q_\mathcal{D}$ is positive. Looking at the equations, a positive $Q_\mathcal{D}$ would act in the same way as a positive cosmological constant, i.e., like Dark Energy. On the contrary, on small scales, the shear term is dominant, since the structures are elongated into a filamentary network of galaxies. The resulting $Q_\mathcal{D}$-term on small scales turns out to be negative. By looking again at the equations this means that $Q_\mathcal{D}$ would add to the density of matter, i.e., it would act like Dark Matter. Hence, the same term can mimic different properties on different spatial scales.

The second new term, $\langle R\rangle_\mathcal{D}$, is the averaged curvature of space: since the curvature is inhomogeneous in the general model, its average $\langle R\rangle_\mathcal{D}$ does not evolve like the curvature in the standard model: the curvature in the homogeneous model and the averaged curvature in the general model have both the same values at every point in space, but the difference comes in with the time-evolution. In the homogeneous model we have $\langle R\rangle_\mathcal{D} = 6k/a^2$, which is the curvature term in Friedmann's equation that does not depend on the region over which we average. In general, however, the behavior of the averaged curvature depends on the domain we look at and its behavior in time is not the same, i.e., in general it does not evolve as $\propto a_D^2$.

Looking at the ensemble of (1.11–1.14) we see that the first is an expansion law like that of Friedmann with the new terms $Q_\mathcal{D}$ and $\langle R \rangle_\mathcal{D}$ added, the second is an acceleration law like that of Friedmann with the new term $Q_\mathcal{D}$ added (this is our (1.2) or (1.3) in the text), and the third equation is again the conservation of energy like in the Friedmann case, however with the averaged inhomogeneous density instead of the homogeneous density.

There is, however, a fourth equation that furnishes another subtle point: in the homogeneous case we said that, if the mass conservation law holds, then the second equation is the time-derivative of the first. The same must be true in the general system. However, since there are these additional terms, they also must be related. And this is the fourth equation that does this for us. This fourth relation is the most interesting in the new cosmological model, since it says that the curvature of space changes as soon as structures form. It corresponds to our (1.1) in the text.

1.8.4 A Compact Form of the New Cosmological Equations

We can rewrite the general cosmological equations (1.11) in the same way as Friedmann's expansion and acceleration laws, however, with different sources:

$$\left(\frac{\dot{a}_\mathcal{D}}{a_\mathcal{D}}\right)^2 = \frac{8\pi G \rho_{\text{eff}}^\mathcal{D}}{3} + \frac{\Lambda}{3} - \frac{k_\mathcal{D}}{a_\mathcal{D}^2}; \tag{1.19}$$

$$\left(\frac{\ddot{a}_\mathcal{D}}{a_\mathcal{D}}\right) = -\frac{4\pi G(\rho_{\text{eff}}^\mathcal{D} + 3 p_{\text{eff}}^\mathcal{D})}{3} + \frac{\Lambda}{3}; \tag{1.20}$$

$$\dot{\rho}_{\text{eff}} + 3\left(\frac{\dot{a}_\mathcal{D}}{a_\mathcal{D}}\right)(\rho_{\text{eff}}^\mathcal{D} + p_{\text{eff}}^\mathcal{D}) = 0, \tag{1.21}$$

where the sources are defined as $\rho_{\text{eff}}^\mathcal{D} = \langle \rho \rangle_\mathcal{D} + \rho_\Phi^\mathcal{D}$ for the actual matter source $\langle \rho \rangle_\mathcal{D}$ and the extra backreaction density $\rho_\Phi^\mathcal{D}$. Notice that backreaction also introduces an effective pressure $p_{\text{eff}}^\mathcal{D} = p_\Phi^\mathcal{D}$. The new backreaction sources are defined in terms of the backreaction variables $Q_\mathcal{D}$ and $\mathcal{W}_\mathcal{D} := \langle R \rangle_\mathcal{D} - \frac{6k}{a_\mathcal{D}^2}$, where this latter is defined as the deviation of the total averaged curvature from the homogeneous curvature term. For the extra backreaction sources we have:

$$\rho_\Phi^\mathcal{D} := -\frac{1}{16\pi G} Q_\mathcal{D} - \frac{1}{16\pi G} \mathcal{W}_\mathcal{D};$$

$$p_\Phi^\mathcal{D} := -\frac{1}{16\pi G} Q_\mathcal{D} + \frac{1}{48\pi G} \mathcal{W}_\mathcal{D}. \tag{1.22}$$

Notice that (1.19) and Friedmann's equations (1.6) are the same (up to the dependence on the averaging domain), if we just replace the homogeneous sources ρ_h and p_h by the effective sources $\rho_{\text{eff}}^\mathcal{D}$ and $p_{\text{eff}}^\mathcal{D}$.

Physically, although the actual matter source contains no pressure, the extra terms coming from the inhomogeneous geometry introduce an effective pressure that can act against gravity. We may now compare with Fig. 1.15. We can conclude that, if we want to describe the Universe with the laws of Friedmann, then there are extra sources that could eventually replace Dark Energy and Dark Matter that had to be introduced in the standard model as being fundamental. We are going to discuss the interpretation of these extra terms as Dark Energy and Dark Matter more carefully below. Before we do so, we mention another way to see these extra terms, and we discuss the example of the equations in the vacuum.

1.8.5 The Morphon: An Effective Scalar Field

In standard cosmology, Dark Energy is often described in terms of a scalar field (so-called *Quintessence*). We said that a positive cosmological constant is the simplest model for Dark Energy, but we may also have a time-varying "cosmological constant". Quintessence models provide such a more general view on the behavior of Dark Energy. Also, there are scalar field models for Dark Matter, where the scalar field can be associated with Dark Matter particles. Models for describing *inflation*, i.e., a period of a very rapid expansion in the early phases of the evolution of the Universe, are also realized by scalar fields. In all these cases a scalar field is added to the sources in the standard cosmological equations of Friedmann and are thought of being fundamental.

We can now look at the generalized cosmological equations above and simply rewrite the extra geometrical terms in the effective sources (1.22) as if they come from a scalar field source. A scalar field source is commonly written in the form:

$$\rho_\Phi^{\mathcal{D}} = \frac{1}{2}\dot{\Phi}_{\mathcal{D}}^2 + U_{\mathcal{D}}(\Phi_{\mathcal{D}}); \quad p_\Phi^{\mathcal{D}} = \frac{1}{2}\dot{\Phi}_{\mathcal{D}}^2 - U_{\mathcal{D}}(\Phi_{\mathcal{D}}). \tag{1.23}$$

This form allows to study the backreaction effect as a mechanical problem of a scalar field $\Phi_{\mathcal{D}}$ rolling in a potential $U_{\mathcal{D}}(\Phi_{\mathcal{D}})$.

By invoking this correspondence, the potential energy density of the scalar field, $e_{\text{pot}}^{\mathcal{D}} = -U_{\mathcal{D}}(\Phi_{\mathcal{D}})$, is determined by the averaged curvature term, and the kinetic energy density of the scalar field, $e_{\text{kin}}^{\mathcal{D}} = 1/2\dot{\Phi}_{\mathcal{D}}^2$, is related to the backreaction term:

$$-\frac{1}{8\pi G}\mathcal{W}_{\mathcal{D}} = 3U_{\mathcal{D}}; \quad -\frac{1}{8\pi G}Q_{\mathcal{D}} = \dot{\Phi}_{\mathcal{D}}^2 - U_{\mathcal{D}}. \tag{1.24}$$

Such a scalar field is called the *morphon field*, since it describes the morphology of the inhomogeneities [6]. Note that for vanishing backreaction, $Q_{\mathcal{D}} = 0$, we fall back onto the standard Friedmann model, and the last equation then characterizes the Friedmann model by the so-called *virial equilibrium condition* $2e_{\text{kin}}^{\mathcal{D}} + e_{\text{pot}}^{\mathcal{D}} = 0$.

Inserting (1.24) into the fourth of the generalized cosmological equations (1.11) then makes the analogy complete: the scalar field has to obey the standard wave equation (the so-called *Klein–Gordon equation*):

$$\ddot{\Phi}_\mathcal{D} + 3H_\mathcal{D}\dot{\Phi}_\mathcal{D} + \frac{\partial}{\partial \Phi_\mathcal{D}} U(\Phi_\mathcal{D}) = 0. \qquad (1.25)$$

We can use this scalar field picture for the backreaction terms to study the same models as those for Dark Energy (Quintessence), Dark Matter, and we can even study the scalar field analogy to classically describe *inflation*, all without adding any fundamental scalar field.

1.8.6 The Equations in the Vacuum

In Sect. 1.5.2 we proposed a "gedanken experiment", where we thought about a four-dimensional vacuum space–time, split into a three-dimensional curved space and a time direction. We can now look at the generalized cosmological equations by putting all matter and energy sources to zero. We are then left with the same equations as they are described above, however, with only the geometrical terms $\rho_\Phi^\mathcal{D}$ and $p_\Phi^\mathcal{D}$ interpreted as sources of a "Friedmannian cosmology". Let us put for simplicity the cosmological constant equal to zero, and let us assume that the fluctuation term $Q_\mathcal{D}$ is very weak on large scales. The expansion law then reduces to:

$$6\left(\frac{\dot{a}_\mathcal{D}}{a_\mathcal{D}}\right)^2 \approx -\langle R \rangle_\mathcal{D}, \qquad (1.26)$$

which shows that, if the model on large scales expands (or contracts!), the curvature has always to be negative on average. The expansion is part of the extrinsic curvature of the space, as it is embedded into space–time. It has to exactly compensate the intrinsic curvature of the space in order to guarantee that the four-dimensional (Ricci) curvature is vanishing (vacuum). (For general estimates on the curvature see [7].)

As explained in the text, this does not mean that the space–time is not curved. What is vanishing is only part of the four-dimensional curvature (the Ricci curvature), since Einstein's equations relate only that part to the sources. The Ricci curvature is just the trace of the so-called Riemann curvature. There is another part, called the Weyl curvature, which is the trace-free part of the Riemann curvature, and this latter can curve up a void, even if we are in an empty world.

We may prescribe initial conditions in such a way that the intrinsic curvature is zero and, hence, the expansion vanishes initially, so that we have split the Ricci-flat space–time into flat and static space sections. However, such a state is unstable, and if we perturb it a little, then the model expands (or contracts) with the consequence of an on average negative spatial curvature.

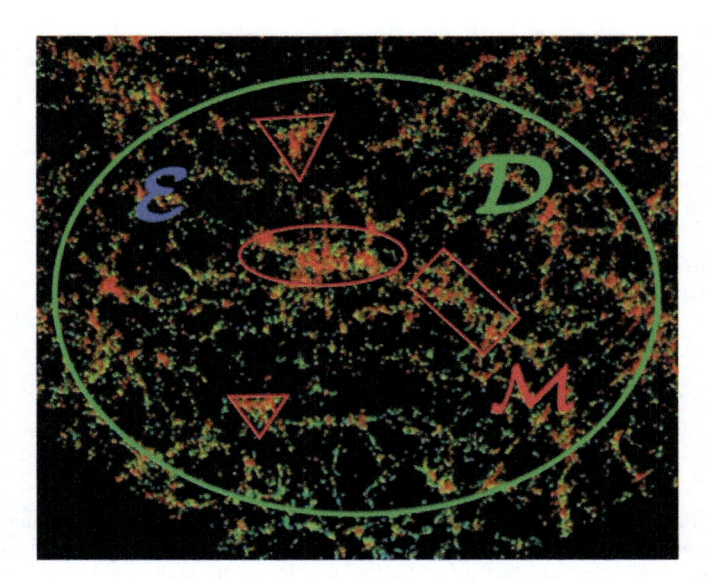

1 Dark Energy and Dark Matter Hidden in the Geometry of Space? 45

Fig. 1.23 A large region $\mathcal{D}$ of the structured Universe is subdivided into an ensemble of overdense regions $\mathcal{M}$ and an ensemble of under-dense regions $\mathcal{E}$. All of the over-densities united in $\mathcal{M}$ and all of the under-densities united in $\mathcal{E}$ make up the total region $\mathcal{D}$. Concrete models for the evolution of this partitioned Universe show that the difference in the expansion rate of the $\mathcal{M}$ regions as compared with the expansion rate of the $\mathcal{E}$ regions results in some sort of "pressure" (stemming from the expansion variance, i.e., the last term in (1.27)) that pushes apart the universe model more faster than if this difference were absent. The standard model is homogeneous and cannot describe such a difference. This "effective pressure" acts like a Dark Energy source that has to be included in the standard model in addition to the ordinary matter density. If we look at the curvature of the different regions, we find that the curvature on $\mathcal{M}$ regions is positive (mimicking a Dark Matter source in the standard model) and the curvature on $\mathcal{E}$ regions is negative. Since the $\mathcal{E}$ regions make up most of the volume, the negative curvature dominates on the large region $\mathcal{D}$ (mimicking a Dark Energy source in the standard model) [Credit: A. Wiegand]

1.8.7 Mimicking Dark Energy and Dark Matter

We said that the backreaction term $Q_\mathcal{D}$ may act repulsively or attracting in the general cosmological equations, so that it could mimic a Dark Energy behavior or a Dark Matter behavior. We also said that concrete models would reveal the former on large scales and the latter on smaller scales. We can make this more concrete by explicitly introducing three sizes of domains (three spatial scales): besides the largest scale of the domain $\mathcal{D}$ we may introduce $\mathcal{M}$ domains that would contain structure with a lot of matter, and $\mathcal{E}$ domains that would contain under-dense regions that are almost devoid of matter. Taking both together makes up the total region $\mathcal{D}$ (see Fig. 1.23). In doing so the kinematical backreaction then assumes the form [7, 21]:

$$Q_\mathcal{D} = \lambda_\mathcal{M} Q_\mathcal{M} + (1 - \lambda_\mathcal{M}) Q_\mathcal{E} + 6\lambda_\mathcal{M}(1 - \lambda_\mathcal{M})(H_\mathcal{M} - H_\mathcal{E})^2, \quad (1.27)$$

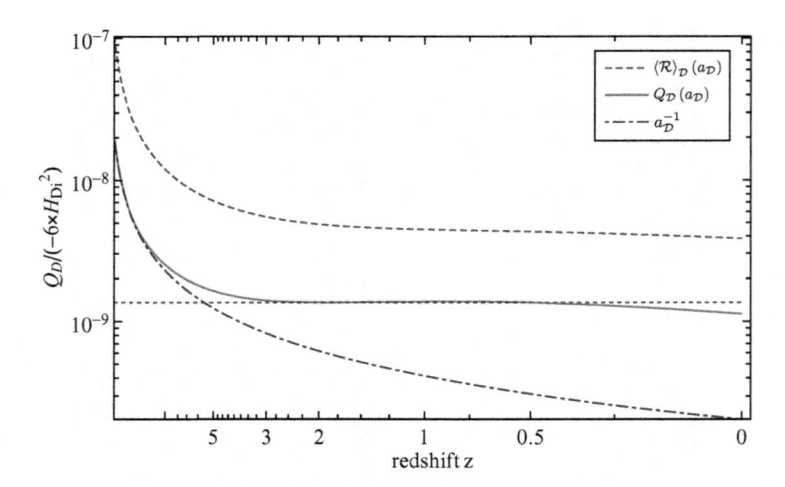

Fig. 1.24 Plot of the evolution of the backreaction term $Q_\mathcal{D}$ and the average curvature $\langle R \rangle_\mathcal{D}$ as a function of the global scale factor $a_\mathcal{D}$. For comparison a line with $a_\mathcal{D}^{-1}$-scaling is added, that corresponds to the assumed behaviors of the backreaction terms on the partitioned domains, and one that is constant. $Q_\mathcal{D}$ and $\langle R \rangle_\mathcal{D}$ are normalized by $-6H_\mathcal{D}^2$ so that the values shown represent $\Omega_Q^\mathcal{D}$ and $\Omega_R^\mathcal{D}$. We appreciate that the backreaction terms feature an approximate cosmological constant behavior on the homogeneity scale $\mathcal{D}$. Physically, this result can be attributed to the expansion variance between the subdomains $\mathcal{M}$ and $\mathcal{E}$ and, hence, this latter is identified as the key-effect to produce a global Dark Energy-like behavior of the scale-factor of the Universe [Credit: Reprinted figure from [21] ©(2010) American Physical Society]

where $\lambda_\mathcal{M} := |\mathcal{M}|/|\mathcal{D}|$ introduces the volume-fraction of the over-dense regions $\mathcal{M}$ compared to the volume of the region $\mathcal{D}$, i.e., it measures how much volume out of the whole volume is dominantly occupied by matter. This fraction may become small in today's Universe, since we see that most of the volume of space is in regions that are almost devoid of galaxies (see Fig. 1.7). Thus, in the beginning of the evolution of the Universe both regions may occupy the same fraction of space, but in the coarse of the clumping of matter into dense clusters, more volume will be in the $\mathcal{E}$ regions.

Putting in the formula (1.27) the backreaction terms $Q_\mathcal{M}$ and $Q_\mathcal{E}$ to zero, there is nevertheless the third term that is positive and depends on the difference between the expansion rates of the different regions: in $\mathcal{M}$ regions the matter is clustered together and no longer partakes in the global expansion, $H_\mathcal{M}$ almost vanishes, while the expansion $H_\mathcal{E}$ in the $\mathcal{E}$ regions contributes mostly to the global expansion $H_\mathcal{D}$. This difference is responsible for an effective pressure that acts like there were Dark Energy: in Fig. 1.24 we discuss the result of a more general model, that also models the backreaction terms on the domains $\mathcal{M}$ and $\mathcal{E}$, to illustrate this fact.

1 Dark Energy and Dark Matter Hidden in the Geometry of Space?

1.8.8 Einstein's Favorite Model: A Static Cosmos

A static Friedmann model is obtained by demanding a constant scale-factor, $a(t) =: a_E$, $a_E = const.$, in (1.6). With this assumption we find $\dot{a}_E = 0$ and $\ddot{a}_E = 0$, and therefore we are left with the following equations (we confine ourselves here to the case without pressure):

$$4\pi G \rho_E = \Lambda = const.; \quad 8\pi G \rho_E + \Lambda = \frac{3k}{a_E^2}. \tag{1.28}$$

The first equation is our (1.4) in the text. Combining these two equations we obtain:

$$\frac{k}{a_E^2} = \Lambda; \quad a_E = \frac{k}{\sqrt{4\pi G \rho_E}}, \tag{1.29}$$

and since $\rho_E > 0$, Λ has to be *positive* and hence also the curvature parameter k.

We see that the cosmological constant Λ is needed, since otherwise Friedmann's equations without pressure would not allow for a static model. This was the reason why Einstein introduced this constant into his equations. Since the curvature has to be positive in the static cosmos, we have a finite spherical space Universe with a finite radius a_E, and a finite volume $V_E = 2\pi^2 c^3 a_E^3$, with the speed of light c. Since the total mass in the Einstein cosmos as well as its volume are constant, we also have a constant density ρ_E. We may then rewrite the above equations as follows:

$$4\pi G M_E = \Lambda V_E; \quad \frac{R_E}{2} = 8\pi G M_E + \Lambda V_E, \tag{1.30}$$

with the total mass $M_E = \rho_E V_E$ and the curvature $R_E := 6k/a_E^2$. ΛV_E may be interpreted as the total *Dark Energy* in this model.

1.8.9 Einstein's Idea in Light of the New Cosmological Equations

In the framework of the new cosmological equations (1.11–1.14) we can closely follow Einstein's idea of a static cosmos. However, here, we construct a globally static model that allows for evolution in its interior. We therefore have to introduce a domain that is identical with the whole Universe, we call this domain Σ. Then, we again require the scale-factor to be constant, but only on the whole Universe $a(t) =: a_\Sigma$ so that the derivatives $\dot{a}_\Sigma = \ddot{a}_\Sigma = 0$. We are left with the following equations:

$$4\pi G \langle \rho \rangle_\Sigma = Q_\Sigma + \Lambda; \quad 8\pi G \langle \rho \rangle_\Sigma + \Lambda = \frac{\langle R \rangle_\Sigma + Q_\Sigma}{2}, \tag{1.31}$$

with the global *kinematical backreaction* Q_Σ, the globally averaged curvature $\langle R \rangle_\Sigma$, and the globally averaged density $\langle \rho \rangle_\Sigma$.

Let us now consider the case of a vanishing cosmological constant: $\Lambda = 0$. We see that here this constant is not needed to construct a static universe model. We find that the backreaction term Q_Σ can play the role of the cosmological constant, and the averaged curvature is, for a non-empty Universe, also positive like in Einstein's model:

$$4\pi G \langle \rho \rangle_\Sigma = Q_\Sigma; \quad 12\pi G \langle \rho \rangle_\Sigma = \langle R \rangle_\Sigma. \qquad (1.32)$$

The first of these equations is our (1.5) in the text.

It is interesting that this cosmos obeys the compact form introduced in Sect. 1.8.4 with the effective sources:

$$\langle R \rangle_\mathscr{D} = 3 Q_\Sigma = const. \Rightarrow p_{\text{eff}}^\Sigma = \rho_{\text{eff}}^\Sigma = 0. \qquad (1.33)$$

The differences to the homogeneous Einstein cosmos are two-fold: first, the property of being static only holds on the entire Universe, but not in the interior, where we have accelerating regions $\mathscr{D}$ with $4\pi G \langle \rho \rangle_\mathscr{D} < Q_\mathscr{D}$, but also decelerating regions with $4\pi G \langle \rho \rangle_\mathscr{D} > Q_\mathscr{D}$. In the former case the region $\mathscr{D}$ would mimic Dark Energy, if the standard model with a cosmological constant were used to interpret this acceleration. The second point concerns the instability of the homogeneous Einstein cosmos: if the balance between the cosmological constant and the matter density is slightly perturbed, it triggers an expansion (the Eddington model) or contraction. In the new globally static model this is not true: if we perturb the balance between the matter density and the fluctuations of matter (the backreaction) slightly, then this would change the curvature in the opposite direction and so again increases or decreases the backreaction with the trend to maintain the balance condition. Finally, we note that in the homogeneous case a non-accelerating model is always non-expanding (static), while in the inhomogeneous case a non-accelerating model may globally expand (or contract) and is not necessarily static.

Acknowledgments It is a pleasure to thank Claus Beisbart, Mauro Carfora, Henk van Elst, Martin France, Tatjana Kumpf, Jocelyne Huguet Manoukian, and Frank Steiner for reading the manuscript and providing valuable remarks.

References

Scientific Articles:
1. R. Aurich, S. Lustig, F. Steiner, H. Then, Indications about the shape of the universe from the Wilkinson microwave anisotropy probe data. "Can one hear the shape of the universe?" Phys. Rev. Lett. **94**, 021301 (2005)
2. R. Aurich, H. Janzer, S. Lustig, F. Steiner, Do we live in a small universe? Class. Quant. Grav. **25**, 125006 (2008)
3. T. Buchert, Dark energy from structure: a status report. Invited Rev. Gen. Rel. Grav. (Dark Energy Special Issue) **40**, 467–527 (2008)
4. T. Buchert, On globally static and stationary cosmologies with or without a cosmological constant and the dark energy problem. Class. Quant. Grav. **23**, 817–844 (2006)

5. T. Buchert, J. Ehlers, Averaging inhomogeneous newtonian cosmologies. Astron. Astrophys. **320**, 1–7 (1997)
6. T. Buchert, J. Larena, J.-M. Alimi, Correspondence between kinematical backreaction and scalar field cosmologies – the 'Morphon Field'. Class Quant. Grav. **23**, 6379–6408 (2006)
7. T. Buchert, M. Carfora, On the curvature of the present-day universe. Class. Quant. Grav. **25**, 195001 (2008)
8. T. Buchert, M. Bartelmann, High spatial resolution in three dimensions – a challenge for large-scale structure formation models. Astron. Astrophys. **251**, 389–392 (1991)
9. N. Cornish, D.N. Spergel, G. Starkman, Circles in the sky: finding topology with the microwave background radiation. Class. Quantum Grav. **15**, 2657 (1998)
10. A.S. Eddington, On the instability of Einstein's spherical world. Mon. Not. Roy. Astro. Soc. **90**, 668 (1930)
11. A. Einstein, Considerations on the universe as a whole. In: Relativity: The Special and General Theory, Methuen & Co Ltd (1920); orig. Über die spezielle und die allgemeine Relativitätstheorie, Gemeinverständlich (1916). Supplement added by Einstein in 1952: Appendix IV: The Structure of Space According to the General Theory of Relativity (The Folio Society, London, 2004)
12. G.F.R. Ellis, Is the universe expanding? Gen. Rel. Grav. **9**, 87 (1978)
13. G.F.R. Ellis, Dark matter and dark energy proposals: maintaining cosmology as a true science? in *Proceedings 2nd CRAL–IPNL conference – Dark Energy and Dark Matter: Observations, Experiments and Theories – Lyon*, July 7–11, 2008, ed. by E. Pécontal, T. Buchert, P. Di Stefano, Y. Copin, France, EAS Publication Series 36, EDP Sciences, (2010)
14. G.F.R. Ellis, T. Buchert, The universe seen at different scales. Phys. Lett. A (Einstein Special Issue) **347**, 38–46 (2005)
15. M. Geller, J.P. Huchra, Mapping the universe. Science **246**, 897–903 (1989)
16. C. Hikage, J. Schmalzing, T. Buchert, Y. Suto, I. Kayo, A. Taruya, M.S. Vogeley, F. Hoyle, J.R.Gott III, J. Brinkmann, Minkowski functionals of SDSS galaxies I: analysis of excursion sets. PASJ **55**, 911–931 (2003)
17. G. Lemaître, Expansion of the universe: a) A homogeneous universe of constant mass and increasing radius accounting for the radial velocity of extra–galactic nebulae. Mon. Not. Roy. Astron. Soc. **91**, 483–490; b) The expanding universe. Mon. Not. Roy. Astron. Soc. **91**, 490–501 (1931)
18. I. Newton, *Philosophiae Naturalis Principia Mathematica* (1713)
19. P.J.E. Peebles, B. Ratra, The cosmological constant and dark energy. Rev. Mod. Phys. **75**, 559 (2003)
20. S. Räsänen, Accelerated expansion from structure formation. Review, JCAP **0611**, 003 (2006)
21. A. Wiegand, T. Buchert, Multiscale cosmology and structure-emerging dark energy: a plausibility analysis. Phys. Rev. D **82**, 023523 (2010)
22. D.L. Wiltshire, Cosmic clocks, cosmic variance and cosmic averages. New J. Phys. **9** (2007)

Popular Articles:
23. G. F.R. Ellis, Patchy solutions. Nature **452**, 13 (2008) *The Universe seems to be expanding ever faster – a phenomenon generally ascribed to the influence of 'dark energy' But might the observed acceleration be a trick of the light in an inhomogeneous Universe?*
24. A. Gefter, Dark energy begone! New Scientist **8**, 32 (2008) *A simple trick of gravity could open the door to a much brighter view of the cosmos.*
25. T. Clifton, P.G. Ferreira, Does dark energy really exist? Maybe not. Scientific American 48, (2009) *The observations that led astronomers to deduce its existence could have another explanation: that our galaxy lies at the center of a giant cosmic void.*
26. J.P. Luminet, G.D. Starkman, J.R. Weeks, Is space finite? Scientific American **280**, 68 (1999)

Internet Sites:
27. http://antwrp.gsfc.nasa.gov/apod/astropix.html Astronomy Picture of the Day
28. http://www.galaxyzoo.org/ Galaxy Zoo Project, where you can help astronomers explore the Universe

29. http://hubblesite.org/ Hubble Space Telescope
30. http://www.ifa.hawaii.edu/~cowie/tts/tts.html Hawaii Active Catalogue of the Hubble Deep Field
31. http://www.sdss.org/ The Sloan Digital Sky Survey: Mapping the Universe
32. http://www.skyserver.org/ The Sloan Digital Sky Server: Galaxy Images
33. http://background.uchicago.edu/~whu/ Wayne Hu's page on the Cosmic Microwave Background
34. http://www.geometrygames.org/CurvedSpaces/ Free Software for Cosmic Topology
35. http://www.projet-horizon.fr/ HORIZON simulations
36. http://www.aei.mpg.de/english/metanavi/links/index.html General Relativity Links, Albert–Einstein–Institut, Germany
37. http://relativity.livingreviews.org/ Living Reviews in Relativity
38. http://www.nasa.gov/centers/glenn/technology/warp/warp.html NASA: Warp Drive Technology
39. http://omnis.if.ufrj.br/~mbr/warp/ Marcello B. Ribeiro: Warp Drive Theory

Chapter 2
The Arrow of Time In a Universe with a Positive Cosmological Constant Λ

Laura Mersini-Houghton

There is a mounting evidence that our universe is propelled into an accelerated expansion driven by Dark Energy. The simplest form of Dark Energy is a cosmological constant Λ, which is woven into the fabric of spacetime. For this reason it is often referred to as vacuum energy. It has the "strange" property of maintaining a constant energy density despite the expanding volume of the universe. Universes whose energy is made of Λ posses an event horizon with and eternally finite constant temperature and entropy, and are known as DeSitter geometries. Since the entropy of DeSitter spaces remains a finite constant, then the meaning of a thermodynamic arrow of time becomes unclear. Here we explore the consequences of a fundamental cosmological constant Λ for our universe. We show that when the gravitational entropy of a pure DeSitter state ultimately dominates over the matter entropy, then the thermodynamic arrow of time in our universe may reverse in scales of order a Hubble time. We find that due to the dynamics of gravity and entanglement with other domain, a finite size system such as a DeSitter patch with horizon size H_0^{-1} has a finite lifetime Δt. This phenomenon arises from the dynamic gravitational instabilities that develop during a DeSitter epoch and turn catastrophic. A reversed arrow of time is in disagreement with observations. Thus we explore the possibilities that: Nature may not favor a fundamental Λ, or else general relativity may be modified in the infrared regime when Λ dominates the expansion of the Universe.

We live in a world that has an arrow of time. Time is a fascinating enigma. We are intrinsically programmed to have a *feeling* of time and of the difference between past and future. Yet there is no consensus as to what the nature of time is and why it has a direction pointing from past to future? St. Augustine described this observation and his view on the nature of time in his book of Confessions: "nevertheless there is time past and future". The echo of these views rings louder in the last decade than over the last two millenia. The renewed interest in the time's

---

L. Mersini-Houghton (✉)
Department of Physics and Astronomy, UNC-Chapel Hill, NC 27599-3255, USA
e-mail: mersini@physics.unc.edu

enigma at present is fueled by two major advances in physics: the discovery of the acceleration of the universe attributed to a dominant mysterious energy, *Dark Energy*; and, a Copernican extension of physics to a multiverse framework. The latter is motivated by the discovery of a landscape multiverse from string theory and from Big Bang inflation believed to be eternally reproducing new universes.

The acceleration of the universe at present is very similar to what happened during Big Bang inflation. The main difference stands in the energy scale at which inflation occurred being 122 orders of magnitude larger than the present energy. The acceleration of the universe can be attributed to a simple constant energy density term allowed by Einstein equation, known as the cosmological constant or vacuum energy. Or in complete analogy with the mechanism driving inflation, a slowly rolling scalar field, mimicking a nearly constant energy density, can be equally responsible as the driving force. Collectively these models are known as Dark Energy. Presently Dark Energy dominates the expansion of the universe by making up for 70% of its total energy budget [2]. The future evolution of the universe is entirely determined by Dark Energy. The universe can continue its accelerated expansion forever, end up in a Big Crunch or, even meet the destiny of a Big Rip whereas the acceleration of the expansion increases with time, resulting in a "tearing" apart of the very fabric of spacetime. Either one of the three options is likely to occur if Dark Energy be dynamic in nature. But if it turns out that Dark Energy is a pure cosmological constant, then the universe continues its accelerated expansion at a constant rate eternally. At the classical level of calculation, in the latter case [3], the universe ends up in a cosmic heat death, cold and empty of structure, maintaining its entropy and temperature at constant values eternally. Such a universe is known as a DeSitter universe.

Let us see what happens to a DeSitter universe when quantum mechanics is taken into account. The second law of thermodynamics states that the amount of disorder, "the entropy", tends to increase. A thermodynamic arrow of time is determined by the direction of entropy increase. When the second law of thermodynamics is applied to the whole Universe, then the immediate implication for the time's arrow is that the Universe must have started in an extremely low entropy state that has been growing ever since its birth to its present value. Low entropy states imply high energies, which are a small subset of the general phase space for the possible initial conditions, a universe could be born with. They are the exception rather than the generic rule. On these basis, Sir Roger Penrose argued that starting the Universe at high energies seems to make the choice of our Initial Conditions very special indeed. Argument about the improbability of high energy states relies on equilibrium statistics, with a possible loophole of eliminating the possibility of some dynamic selection being at work. Statistical mechanics estimates, often used in literature, can quantify the Penrose statement as follows: the probability to have an initial patch inflate at some high energy scale Λ_i goes as $P \simeq e^{S_i}$ where the entropy $S_i = 3/\Lambda_i$. This expression indicates that a GUT scale inflating patch like ours, is the most special and unlikely event to have started the universe, as likely as 1 part in $10^{10^{123}}$. Yet without this low entropy we can't explain the observed arrow of time.

In [14] we proposed a new approach, based on the quantum dynamics of the gravitational degrees of freedom (DoF), for the investigation of the Initial Conditions. We showed how the dynamics of the combined, matter + gravity system, is out-of equilibrium and how this dynamics breaks ergodicity. The compression of the phase space into a small region of initial conditions corresponding to low entropy and high energy inflation resulted in the emergence of a new superselection rule for Initial Conditions with very different probability estimates from the previous arguments stated above. The superselection of high energy patches from the non-ergodic phase space helped us to understand why the Universe had to start from extremely low entropies, thereby with an arrow of time.

Let us briefly highlight the dynamics of the combined system below as it becomes relevant again for the late time accelerating Universe. The basic mechanism that picks out the high energy "corner" in the phase space of initial states is the following: gravitational DoF, here corresponding to the degrees of freedom on the DeSitter horizon $H_0 \simeq \sqrt{\Lambda}$, comprise a "negative heat capacity" system. Therefore gravity tries to reach equilibrium by tending to expand the spacetime to infinity; matter is a "positive heat capacity" system and thus it goes towards equilibrium by collapsing; the dynamics of the combined system, matter + gravitational DoF, is an out-of equilibrium system with two opposing tendencies "fighting" to expand or collapse the patch. Depending upon which one wins, it can result in a collapsing initial patch or a growing one, thus creating irrelevant Initial Conditions for starting a universe when the initial state can not grow. Or, if the expansion driven by the dynamics of the gravitational DoF's "wins" over the crushing effect of the dynamics of matter DoF's, the results is an expanding initial patch which result in a physically relevant Initial Condition that gives birth to a universe. We belong in the latter, the "survivor universes" case. Quantitative, the entropy of the "survivor universes" is given by the Euclidean action of the system which is modified in the presence of gravity. In fact the modified entropy is even less than its equilibrium expression, due to the out-of equilibrium dynamics. Entropy is given roughly by log of the volume in phase space. The modified entropy is then a direct consequence of the reduction of the relevant phase space to a small "corner" of its initial volume that contains the "survivor universes" only.

It is possible that our Universe may be going through inflation again at late times due to the domination of Dark Energy in its energy budget. If true, then the dynamics of the gravitational DoF is expected to play a major role again at late times and a quantum treatment of DeSitter spaces becomes relevant. Within the framework of our theory for the dynamically selected birth of our universe from a multiverse, we would like to ask: What is the ultimate state of such a Λ-universe and can it achieve equilibrium?

Current astrophysical observations provide convincing evidence that our universe is indeed accelerating again at recent times, this time at an energy scale $\Lambda \simeq (10^{-3}\text{eV})^4$. Determining whether a pure vacuum energy Λ or some dynamic mechanism is at work is of paramount importance and intrinsically related to understanding and predicting the future evolution of our universe. But, a well known problem in cosmology is that the outcome of cosmological observables is observer dependent. Thus we need to define our observer. A key point to keep in mind is that in order to *meaningfully* ask questions about the selection of our Initial and Boundary Conditions, quantum numbers or constants of nature, we need to rely on a super-observer, defined as the one that resides in the multiverse of the landscape or equivalently in the allowed relevant part of the phase space. In short we need an "out-of our box" observer in order to compare possible choices [15, 26, 32]. A local internal observer, defined as the one bound to our casual patch (the volume inside the horizon is the "box"), is always surrounded by the horizon thus it can never observe or be aware of the existence of other patches outside the horizon. It is impossible for the internal observer to ask questions about the selection of our initial and final conditions and our constants of nature because the observer is forbidden to ever know that possibilities other than the universe it observes, may exist. Using the long wavelength perturbations of inflation as the super-observer, let us now explore the implications of the proposal in [14] for the late time acceleration of the universe.

We show below how the final destiny of a DeSitter universe is to become quantum again, roughly in times of order a Hubble time. While in real spacetime universes are described by their spatial and time dimension, in the abstract phase space they are represented by localized quantum wavepacket that flow along well defined paths. These paths known as classical trajectories are determined by equations of physics and the decoupling among the various paths (disentanglement of wavepacket universes from each other) is achieved via a process known as decoherence that ensures the quantum to classical transition of the universe at birth. But should the wavepacket reach a turning point in its classical path, then the universe starts displaying its quantum nature again. We show, that is, what happens to a cold and empty DeSitter universe in which quantum fluctuations are the only mechanism left. As the universe becomes quantum thereby reverses the thermodynamic arrow of time at the turning point of its classical trajectory. We will refer to the transition to a quantum state as recoherence of the wavepacket. The details of the quantum state near the turning point depend on the boundary conditions chosen for the wavefunction which are important for the puzzle of the agreement between a thermodynamic and a cosmological arrow of time. We do not embark upon this puzzle here. The key point in this investigation is the fact that the Universe becomes quantum while it is trying to reach the thermal equilibrium of a pure classical DeSitter state. As a result it never reaches this equilibrium. We investigate how the universe transits into the quantum state when it approaches a pure DeSitter state and how this transition affects the thermodynamic arrow of time. We argue on these basis that within the framework of general relativity as the theory of gravity, nature may not favor the existence of Desitter spaces, i.e., of a pure fundamental constant Λ. We then speculate in the discussion section that should such an exclusion principle exist in nature, it would leave us to conclude: either the possibility of a temporary bleep of acceleration due to some dynamic [28] or transplanckian [27] mechanism or; to the consequence – *if we must have Λ then perhaps a new theory of gravity in the infrared regime could be taking over and replacing general relativity* [29].

2.1 The Problem with the Thermodynamic Arrow of Time in DeSitter Spaces

Suppose we do have a new fundamental scale at low energies in our Universe, the DeSitter horizon $H_0 \simeq \sqrt{3/\Lambda}$, given by the pure vacuum energy $\Lambda \simeq (10^{-3}\text{eV})^4$. Observations will soon pin down the equation of state w for Dark Energy, thus distinguishing between a pure Λ and a dynamical mechanism. Observationally we also know and agree that we have an arrow of time, pointing from past to future, determined by the direction of the entropy growth. What does the constant horizon entropy of the final state of the ΛCDM universe imply for the arrow of time and ultimately for the DeSitter universe?

Based on the assumption that Dark Energy is a fundamental Λ, it seems that we ultimately must end in a DeSitter space. The gravitationally driven expansion ultimately wins [3] for the following reason: in contrast to the entropy of DeSitter space which remains eternally a finite constant, not only does matter dilute with time,[1] but also as we accelerate, parts of our spacetime lose causal contact with each other, swallowing matter with them. Ad infinitum we lose all the matter from our universe in this manner and the entropy contribution associated with it. Ultimately the only entropy component left is the gravitational one associated with the DeSitter horizon. Unlike Black Holes, a DeSitter horizon does not evaporate away. Matter provides the environment that ensured decoherence. In terms of wavepackets, the entanglement among various wavefunction is described by the off-diagonal terms in their possible products, known as the density matrix. Thus the Gaussian suppression in the product of different wavefunctions, with width Ω_R^{-1} in the off-diagonal elements of the density matrix ρ, ([4, 6, 16], see also (2.12)), describes how fast our (patch) wavepacket vacua ϕ decoheres from the other wavefunctions and becomes classical, $\rho \simeq e^{-\Omega_R a^4 (\phi - \phi')^2}$. All the other wavefunction in the multiverse provide the environment with which our universe can entangle. Decoherence and the appearance of a classical world comes about from the backreaction of the environment on our branch of the wavefunction $\Psi(a,\phi)$. Losing causal contact with matter seems to imply that we reach thermal equilibrium by evolving in a pure DeSitter state since we lose the environment and the coarse-grained entropy growth provided by it. This reasoning implies that the energy level in the path in phase space for our DeSitter universe should broaden. Since the DeSitter horizon H_0 classically is a constant then at the classical level so is its entropy, constant and finite. But the thermodynamic arrow of time is determined by the entropy growth. At the classical level, it is not clear at all what happens to the arrow of time during an eternal DeSitter epoch when its classical entropy has to remain a constant. These paradoxes indicate that we must search for an answer not at the classical level but by investigating the quantum dynamics of the gravitational degrees of freedom of the system.

---

[1](e.g., in Black Holes by radiating away).

We showed in [14] that during an inflationary phase higher multipoles develop gravitational Jeans-like instabilities which illustrates the importance of the dynamics of gravitational degrees of freedom. The entropy of the universe is lowered in the combined matter + gravity inflating system. At early time, only patches inflating at high energies Λ_i, survived this instability. We discussed in [14, 15] how the mixing in the initial state must survive at late times, due to unitarity (see also [13]). The reduced density matrix obtained by integrating out the environment and superHubble matter modes, gave a scale of mixing and decoherence for our patch at initial times. It also provided us with the evolution equation of this mixing/coherence length at late times namely, $[\hat{H}, \rho] = i\frac{d\rho}{dt}$ where $\hat{H}$ is the Hamiltonian operator, values of which describe the energy of the system.

The initial decoherence width described, how fast the quantum to classical transition for our universe occurred at early times. The fact that the density matrix evolves with time $[H, \rho_r] \neq 0$, [14,15] illustrates why, with gravity switched on, we should not be tempted to use equilibrium expression like $P \simeq e_E^S$ again, and why we can not use causal patch observers for investigating this problem. The universe is in a mixed state and out-of equilibrium even at late times. How can it evolve to a pure state to reach equilibrium?

Classically, the ultimate destiny of the Λ universe is a cold DeSitter space in thermal equilibrium, with finite final entropy $S \simeq 3/\Lambda$, a pure state. Can an ultimate pure state be allowed for the DeSitter space? The environment in our investigation of the dynamics of matter + gravity DoF, is made of the matter modes within our casual patch as well as the backreaction of higher multipoles with superHubble wavelengths. By losing matter inside our casual patch means we are losing part of the environment, which is very important in providing the Gaussian suppression for the off-diagonal elements of ρ_r that ensure decoherence and a classical world. Gravity winning over matter seems to imply that we are evolving into a pure state, rather than a mixed state on the landscape phase space. *But, if we assume a unitary evolution, we can not evolve in a pure state.* Perhaps, even with matter gone, some degree of mixing must remain. This entanglement can come only from interaction with superhorizon modes, i.e., from "out-of-the box". In fact it does. In the DeSitter stage, the environment comes from the entanglement and interaction of our DeSitter patch with the higher multipoles [14]. The dynamics of this mixed state is highly nontrivial and we set to explore it below.

Many authors have discussed the discrepancies and apparent paradoxes that arise from a finite DeSitter entropy in the final state of the universe. In [7–9, 12] authors point out that a finite entropy implies a finite size Hilbert space which is incompatible with DeSitter symmetries $SO(d, 1)$. Meanwhile in [21] by geometric arguments, authors realized that DeSitter space may be unstable, in accord with our results in [14, 15]. Authors of [8] then argued that perhaps DeSitter state is metastable on the landscape and it either goes through a Poincare cycle or tunnels to lower energy vacua. The decay time for tunneling to lower states T_f seems shorter than the recurrence time T_r, $T_f \simeq e^{S_o - S_f} \ll T_r \simeq e^S$, where S_o is the vacua entropy and S_f the fluctuation entropy that would take the state to the top of the potential and allow it to roll the other way into the lower energy vacua. This picture fits well with eternal inflation where many pocket universes at different energies are created and transition among them is possible. The possibility that a DeSitter state is not eternal but rather metastable due to tunneling to lower energy states, is certainly plausible.

But, even if it sounds unlikely, what if Λ is a globally fundamental scale, i.e., suppose that this DeSitter state has no other lower state to tunnel to? What then?

Let us show that even without tunneling, a DeSitter state is not just metastable, but gravitationally it is a catastrophically unstable state. We show here that even before it becomes metastable from tunneling, the *growth of gravitational instabilities* force the DeSitter patch to reverse its arrow of time by hitting a turning point in its trajectory in phase space. We show this phenomenon occurs in time scales of order roughly horizon scale, $T_{\text{crunch}} \simeq \sqrt{\Lambda}$ which is exponentially shorter that tunneling time T_f. This result would be in a huge disagreement with observations. We do not observe a reversal of the arrow of time in our DeSitter universe, in time scales of the order of the age of our universe.

2.1.1 Big Crunch Through Recoherence or a New Exclusion Principle?

Suppose we are in the far future where we have lost almost all matter contact from our casual patch. As we discussed above, the only environment we are left with now comes from the backreaction of the superHubble matter modes, i.e., matter fluctuations of order DeSitter horizon H_0 or larger. We would like to know the evolution of the wavefunction for our universe, and the evolution of the density matrix. The wavefunction of the universe propagates on a minisuperpsace defined by the variables: ϕ_k that labels the vacua with energy E_k on the landscape multiverse; $a[t]$, the scale factor that describes the geometry of the universes; and, f_k that describes the perturbation modes around each vacua, with wavelengths larger than the horizon. The density matrix describes the effect of the interaction of superhorizon modes on our DeSitter geometry. Proceeding as in [14], let us assume that within the WKB approximation, the wavefunction of the universe is given by, $\Psi[a, f_n, \phi] \simeq e_E^S$, $\Psi = \Psi_o \Pi_n \psi_n$, $S \simeq S_0 + \Sigma_n S_n \frac{1}{2} f_n^2$ where S_E is the Euclidean action equivalent to the entropy, and fluctuations f_n correspond to an expansion around zero of the matter field $\phi = \phi(0) + \Sigma_n f_n(a) Q_n(x)$ which can be written as

$$\Psi \sim \Psi_0(a, \phi_0) \prod_n \psi_n(a, \phi_0, f_n). \tag{2.1}$$

The Wheeler–DeWitt Master Equation, that includes the backreaction $\hat{H}_n$ of the matter higher multipoles $f_n{}^2$ is obtained from the quantized Hamiltonian

$$\hat{H} = \hat{H}_0 + \sum_n \hat{H}_n \tag{2.2}$$

---

[2](obtained by the expansion around zero of the landscape minisuperspace field ϕ, see [14, 15]).

acting on the wavefunction, [4,5]

$$\hat{H}_0 \Psi(a,\phi_0) = \left(-\sum_n \langle \hat{H}_n \rangle\right) \Psi(a,\phi_0), \tag{2.3}$$

where the angular brackets denote expectation values in the wavefunction ψ_n and $\hat{H}_n$ denotes the backreaction correction from interaction with the superhorizon fluctuations f_n. The potential term of the backreaction corrections during a nearly DeSitter phase [4, 14], calculated from the inhomogeneous matter perturbations in [5], to second order, is

$$U_\pm(\alpha,\phi) \sim e^{6\alpha}\left[\frac{n^2-1}{2}e^{-2\alpha} \pm \frac{m^2}{2}\right],$$

with $m^2 \simeq \sqrt{\Lambda}$ at present and $U_\pm$ sign corresponding to the interaction coming from subHubble(+) or superhorizon(−) multipoles respectively, as summarized in the appendix and shown in Fig. 2.3. During the DeSitter stage the backreaction energy density $< U f_n^2 >$ becomes negative, $U = U_-$, since our environment contains only the modes with wavenumber $n \leq aH_0$, i.e., with superhorizon wavelengths [6, 14]. The index 0 refers to present day values and $H_0 \simeq \sqrt{\Lambda}$ is the present DeSitter Hubble constant, with a the scale factor. This negative contribution yields an oscillatory correction to the action and it gives rise to tachyonic fluctuations in (2.3). The backreaction Hamiltonian, during the DeSitter stage with curvature Λ, is roughly

$$2e^{3\alpha}\hat{H}_n = -\frac{\partial^2}{\partial f_n^2} + e^{6\alpha}\left(-\sqrt{\Lambda} + e^{-2\alpha}(n^2-1)\right) f_n^2, \tag{2.4}$$

with $\alpha = \text{Log}[a]$, $\hat{H}_n \simeq \Sigma_n U_n f_n^2$ where matter fluctuations $< f_n^2 > = O(H_0)$. During the matter dominated phase the superhorizon modes are nearly frozen thus their kinetic term contribution can be ignored. During the DeSitter phase with vacuum energy Λ, our environment consisting of superhorizon matter fluctuations, contributes a negative backreaction, $f_n^2 U_n = -(\sqrt{\Lambda} - n^2/a^2)f_n^2$, as can be seen from the expression that $U_n < 0$ is negative for the superhorizon higher multipoles $n \leq aH_0$, [6, 14]. This negative contribution drives the gravitational instability of the massive superhorizon size modes f_n, that is, during the DeSitter stage f_n modes start growing exponentially with time, (see the appendix and [14] for the f_n equation of motion).

The correction term Ψ_n to the wavefunction satisfies the following equation:

$$i\frac{\partial \Psi_n}{\partial t} = \hat{H}_n \Psi_n, \tag{2.5}$$

which for the modes $n \leq aH_0$ with H_0 our present DeSitter horizon, and $U_n < 0$ gives a solution of the form, $\Psi_n \simeq e^{-\frac{1}{2}f_n^2 S_n(a)}$ where, the WKB 2nd order correction term to the action is $S_n \simeq i \int a^3 U_n \frac{d\ln(a)}{H}$. Since $U_n < 0$ and S_n is not real in this regime, then this correction breaks the finiteness condition.[3]

Perhaps a better way to see the role of the superhorizon size nonlocal entanglement of our DeSitter patch with f_n, or equivalently in the phase space description, the effect of the backreaction term on the classical path of the wavefunction is by integrating out the Friedman-like equation which includes the entanglement term. This equation is obtained from the generalized Hamiltonian in the Master equation (2.2), which contains the backreaction energy, $\Sigma_n \hat{H}_n$

$$3M_p^2 \dot{a}^2 - \Lambda a^2 + aH_0^3 = 0, \qquad (2.6)$$

where $\Lambda = 3M_p^2 H_0^2$. The last term comes from the backreaction of the higher multipoles $k = n/a \leq H_0$ thus it is negative during the DeSitter stage as explained above. A rough estimate is obtained by summing over all the superhorizon modes, $\Sigma_{aH_0}^0 < \hat{H}_n >$, where $\Sigma_{k=n/a=H_o}^0 (n/a)^2 dn f_n^2 \simeq aH_0^3$. (Below we take $8\Pi G_N = 1$ unless explicitly stated). From (2.6) we can see that our classical trajectory reaches a turning point in phase space, $\dot{a}_* = 0$, during a finite time a_* given by:

$$a_*^2 \Lambda = a_* H_0^3, a_* = \sqrt{\Lambda}. \qquad (2.7)$$

This means that due to the gravitational instability developed by the tachyonic superHubble modes, $U_n < 0$, our DeSitter space goes through a crunch in real spacetime or a turning point in phase space in a time given by the condition $a_* = O(H_o)$. Solutions to (2.6) allow us to estimate the time, t, that the transition takes to go, from the regime of a classical large Λ dominated DeSitter universe with scale factor a and Hubble expansion, H_0, to the quantum regime of the Universe as it approaches the crunching turning point, where the scale factor, $a \to a_*$ becomes very small and the Hubble expansion rate, $\dot{a} \to \dot{a}_* = 0$ approaches zero. This solution is

$$a(t) = \left(\frac{H_0^3}{\Lambda}\right) \text{Cosh}^2 \left[\frac{1}{2} H_0 (t - t_*)\right]. \qquad (2.8)$$

Clearly when $t \to t_*$ we have $a \to a_*$, i.e., The universe's scale factor is reducing to a very small size and the turning point is approached at time t_*. Prior to the instabilities growth, in the regime where the universe is a classical Λ dominated DeSitter space, i.e., for $t \ll t_*$, we see that the scale factor a is nearly exponentially large, and corresponding to that of a large DeSitter space. *We can readily estimate*

---

[3] We can treat the contribution of each mode f_n separately since they are decoupled from each other.

from our solution, (2.8) that the time $\Delta t = (t - t_*)$ *the Universe takes to transit from a large classical regime with size given by the Hubble radius,* $a = r_{H_0}$, *to a small quantum regime* $a = a_*$ *is* $\Delta t \simeq \frac{122}{H_0}$, *roughly about* 122 *times the DeSitter Hubble time*. This is an amazing result, nothing short of an uncertainty principle! In this work, by taking into account the quantum dynamics of gravity, (2.7), leads us to an uncertainty principle for our quantum cosmological "intrinsic clocks" given by $a(t)$ for the DeSitter system with size H_0^{-1} namely, $a(t) H_0^{-1} \simeq l_p^2$ where Planck length l_p is unity in Planck dimensions. Using the solution for $a(t)$ in (2.8) this result can be translated into an uncertainly principle for the physical time, t, of the DeSitter patch, given by $\Delta t H_0 = \Delta N \simeq \text{Log}[M_p^4/\Lambda]$ where ΔN is the number of efoldings that have crossed the horizon since the inflationary time, i.e., defined by $a = e^N$. It is reassuring that although DeSitter space is eternal at the classical level, our findings for a catastrophic collapse of DeSitter space at the quantum level during the time interval Δt due to nonlocal entanglement and the dynamics of gravity, are consistent with the uncertainty principle. Furthermore, the interpretation we offer above for the nonlocal entanglement of the superhorizon modes f_n performing a quantum measurement on the DeSitter patch is now justified by this uncertainty principle. (Note that in the above (2.6) we have not included the matter and radiation contribution to the energy density of the Universe, as we are interested here in investigating the transition regime when the Universe is becoming a DeSitter one, i.e., from the time when Λ dominates over the other components, and onwards. Subsequently the solution found for $a(t)$, (2.8), reflect that fact through the nearly exponential time dependence, the Cosh^2 term.).

Since our classical trajectory $\phi(a)$ goes through a turning point induced by the gravitational instabilities of backreaction in the DeSitter epoch, $\dot{\Psi} = 0$, it follows that after losing casual contact with the internal environment, the universe becomes quantum again as we approach the turning point in a time a_*. If observers bound to the causal patch survived while the universe is transiting to a quantum state, they would perceive the arrow of time as being reversed near the turning point. However note that very close to the turning point, the WKB approximation breaks down, therefore the concept of time and any statement about what could be observed becomes very fuzzy. A better estimate as to when effects related to "recoherence" display a significant effect in our universe, is obtained from the density matrix below, (2.11). Physically, the conditions placed on the wavefunction near its turning point, imply a mixing of "future" and "past" branches of the trajectory. That is, somehow the universe is "aware" of its future evolution since the forward and backward direction contribute into making the wavepacket of the universe. Near the turning point, the two branches merge resulting in a zero speed for the wavefunction there, (thus the name "turning point"), Fig. 2.1. The behavior of the wavefunction can not be described classically anylonger and the intrinsic quantum nature of the wavefunction takes over as it approaches the turning point.

We would like to explore the evolution of the density matrix ρ and the Gaussian suppression width, in order to understand better the effect of the interaction of superHubble modes with the DeSitter geometry and the (re-)appearance of a quantum world during the reversal of the arrow of time near the turning point.

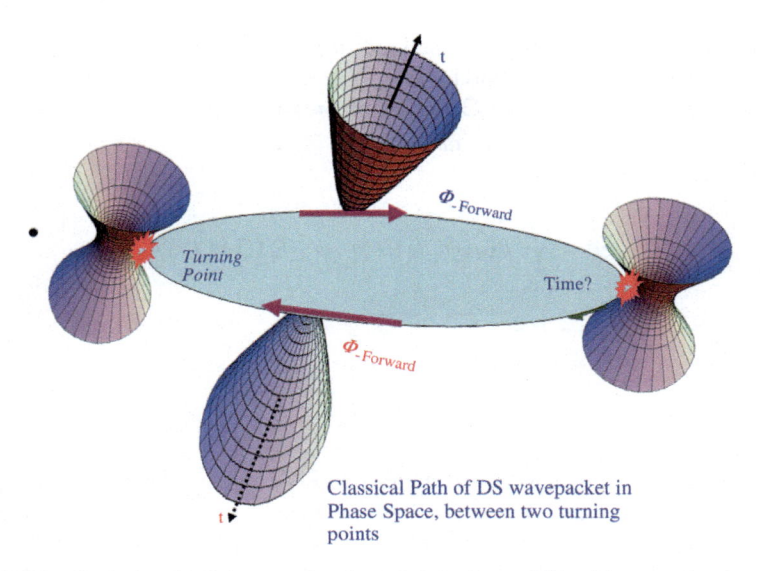

2 The Arrow of Time In a Universe with a Positive Cosmological Constant Λ

Fig. 2.1 The classical path of the wavefunction of the universe $\Psi[\phi, a]$ between its two turning point, in phase space. The wavepacket is a mixture of the forward and backward branches in order to have a vanishing speed near its turning point, $\dot{\Psi} = 0$

We exhibited in [14] that the phase space is not conserved due to these interactions. The equation that governs the evolution of ρ is

$$[\hat{H}, \rho] = i \frac{d\rho}{dt}. \tag{2.9}$$

After exit from inflation, during a matter dominated universe, the matter environment within our casual patch provides a strong suppression in the off-diagonal elements of ρ [16], with width given by Ω_R^{-1},

$$\rho \simeq e^{-\Omega_R a^4 (\phi - \phi')^2}. \tag{2.10}$$

(Note that a large Ω_R means high squeezing in ϕ but a broad wavepacket in the conjugate momenta p_ϕ of ϕ). We can now obtain the evolution of ρ by solving (2.9) for late times, when the small vacuum energy Λ starts dominating the total energy density in the Universe, $\hat{H}$, by including the effect of backreaction term. Solutions to (2.9) give the following:

$$\rho \simeq e^{-\left[\frac{\Omega_R}{a^2} + \mu^2\right] a^6 (\phi - \phi')^2}. \tag{2.11}$$

The second term is the correction coming from the backreaction correction, S_n, where $\mu^2 \simeq < \frac{n^2 a}{H_0}[1 - \frac{1}{3}(\frac{m^2 a}{n})^2] f_n^2 >$. The first term is much studied in literature, (see, e.g., [16]). For the sake of illustration, we can use as an approximate estimate for the first term Ω_R the results obtained in, e.g., [17]. Clearly, $\mu^2 \leq 0$ for the backreaction of superhorizon modes $n \leq \sqrt{2/3}aH_0$, on the DeSitter Universe. We can therefore see that the total Gaussian suppression in (2.11), $\xi = [\frac{\Omega_R}{a^2} + \mu^2]$ is decreasing and approaching zero fast as $\mu^2 \simeq \Sigma_n S_n f_n^2$ becomes negative.

The reduced density matrix:

$$\rho = \int \Psi(a, \phi, f_n) \Psi(a', \phi', f_n') \Pi_n df_n df_n'$$
$$\simeq \rho_0 e^{-\frac{(a\Pi)^6 H^2 \mu^2 (\phi-\phi')^2}{2}}$$
$$\rho_0 \simeq \langle \Psi_0(a,\phi) \Psi_0(a'\phi') \rangle$$
$$\simeq \rho(a,a') e^{-\xi a^6 (\phi-\phi')^2}. \tag{2.12}$$

Note that since $\mu^2 < 0$, we effectively have $\xi = \frac{\Omega_R}{a^2} - |\mu^2|$. As $\xi \to 0$ the classical trajectory in phase space ϕ is spreading thus losing its classicality and becoming highly quantum. The increasing width ξ^{-1} of the Gaussian cross-term of the density matrix ρ describes the broadening of the wavepacket during the transition to a quantum universe. What effects would we observe as the universe is going towards the transition? This is a difficult question since near the turning point, the width of the Gaussian suppression goes to infinity, thus the WKB approximation breaks down at the turning point. Hence, if we were to speculate as to what would we observe away from the turning point where the semiclassical treatment is still valid, as the broadening of the wavepacket is becoming comparable to the difference between neighboring energy levels ϕ, ϕ', then we would probably expect to see an inhomogeneous growth of anisotropies in the sky and probably different values of Λ and quantum numbers in different parts of the sky, i.e., the opposite of the decoherence effect in the early universe [11].

Varying the above action of matter + gravity, $S = S_g + \Sigma_n S_{n,\text{backreac}}$ and using the $\phi(a)$ trajectory equation of motion, we can obtain the turning point for ϕ by the condition $\xi = 0$ which gives $a_* \simeq \sqrt{\frac{2\Lambda}{3}}$ using the expressions above for μ and for the time evolved solution ρ. During a DeSitter phase in a time a_*, the Gaussian width that suppresses the off-diagonal element of ρ_r in ϕ space is increasing due to the developing instability, $\rho_r \simeq e^{-[\Omega_R - a^2\mu^2]a_*^4(\phi-\phi')^2}$. The energy level in phase space is broadening in the landscape coordinate ϕ. By the uncertainty principle a broadening in ϕ means high squeezing in the space of its conjugate momenta p_ϕ since $\Delta\phi\Delta p_\phi \simeq 1$. Thus the big crunch in real space and the quantum recoherence in our universe as it goes through its turning point in phase space. We can estimate the entropy of the DeSitter universe as it is heading through its turning point by noting that the Euclidean action is the entropy of the system. Roughly, $S \simeq |r_{\text{DeSitter}} - r_{f_n}|^2 \simeq (\sqrt{\frac{3}{\Lambda}} - \frac{3a^2}{2\Lambda^{3/2}}) \to 0$. This means that the total entropy for the

DeSitter state including its horizon size fluctuations goes to zero at $a_* \simeq O(\sqrt{\Lambda})$,[4] as the universe becomes quantum through its turning point in phase space and big crunch in real space. This is to say that the allowed volume in phase space for this classical configuration shrinks to a point, i.e., nearly zero. And so is the dimension of the Hilbert space for it. This implies that there are no states available in the Hilbert space for a DeSitter universe.

2.1.2 What Have We Learned?

The analysis carried in this work illustrates the fact that superhorizon nonlocal entanglement and the dynamics of gravity can not be ignored during a DeSitter era. When the DeSitter Universe loses causal contact with the internal environment, its only interaction left is with the higher multipoles. During this stage the DeSitter patch recoheres and ends up with its quantum entropy, which is zero, due to developing Jeans-like instabilities. An immediate consequence of this statement is that the arrow of time is reversed within a time a_* during the DeSitter phase, due to the appearance of coherence and broadening of levels, as the Universe heads towards its gravitationally induced turning point. Although the universe becomes quantum by being superimposed to its own reversed copy and other patches at its turning point, we as internal observers perceive this reversal as a catastrophic crunch in our patch. The details at the turning point depend on the boundary conditions imposed on $\dot{\Psi}(a_*)$. However at the turning point the quantum universe is superimposed to its own time reversed copy and possibly other patches such that its momenta in phase space squeezes to nearly zero. The instability time when this phenomena occurs is, as we have shown here, $a_* \simeq O(H_0)$. This time scale is exponentially shorter even than the tunneling time, if the metastable DeSitter space could lower its energy by decaying. Tunneling to other vacua may not help save the DeSitter Universe. Due to the quantum dynamics of gravity this unstable state crunches catastrophically by becoming quantum during the time scales that it would have reached its final equilibrium and settle into a pure state. The mixing does not allow DeSitter space to evolve and settle into a classical and pure equilibrium state with constant entropy at the end. Instead as the universe is trying to reach its DeSitter equilibrium, this entanglement induces a violent transition to a quantum universe for the final state of the DeSitter space. This recoherence occurs much before the universe can tunnel through to a lower energy configuration. As the DeSitter universe is becoming unstable, internal observers in this patch, perceive their thermodynamic arrow of time as being reversed during a time scale a_* of order a Hubble time. Clearly we

---

[4] A result we could have expected physically, for a quantum DeSitter state but that is confirmed here by a different calculation.

don't observe this in our universe. This fact in combination with the result that the volume of phase space for DeSitter states is zero leads us to conclude: Nature forbids the existence of a pure cosmological constant.

2.2 Comments

Then what are we going to make of the observed acceleration of our universe? Based on the above analysis, we would predict that this acceleration is a temporary bleep due to some dynamic or transplanckian field and not our final destiny due to a fundamental scale Λ in nature. Observations will soon tell us whether this is true or false. In our analysis we assume general relativity as the theory of gravity and the validity of quantum mechanics. If it turns out that we observe a pure Λ then this must signal the breakdown of the general relativity in the IR regime and the emergence of a new theory of gravity by which our arguments above would not hold. It is though entirely possible that, just like the second law of thermodynamics, we may have a new empirical law in nature, let us call it the "Entropy Exclusion Principle", that is: *Nature May Not Allow a Fundamental Λ*.

How can these predictions be tested? The combined observations from SN1a, large scale structure LSS and CMB, from existing or upcoming experiments like Planck, SNAP and LISA, will soon be able to pin down whether the equation of state of Dark Energy is a constant or if it evolves with time [24]. We concluded here that if it turns out we do observe a cosmological constant then gravity must be modified in the ultralarge scales, the IR regime. A modified theory of gravity would leave a strong signature on large scale structure of the universe [22,23]. Dark Matter experiments, e.g., [25] and weak lensing tomography for LSS would reveal such deviations and test our predictions.

It is interesting to note that the initial and boundary states are selected by the same superselection rule in phase space. Without a precise knowledge of quantum gravity and the phase space for the wavefunctions, we can only speculate as to what happens next: if quantum mechanics remains valid deep into the quantum gravity realm then, after the turning point crunch, for example with a boundary condition $\dot{\Psi} = 0$, the "DeSitter" universe would travel its reversed path in the trajectory towards its initial point,[5] therefore completing one of its cycles between initial and final states. Such a condition on the wavefunction also implies a mixing of past and future, of

---

[5]For a different choice of boundary conditions for the wavepacket in phase space, the reversed path can be very different from the forward one depending on what WKB wavefunctions in the wavepacket dominate the action in the reversed direction, in which case there may not even be a cyclic behavior at all. Subtleties related to the boundary conditions at the turning point were investigated by Hawking and Page [30] for the case of closed universes, pointed out to me by R. Holman. It would be interesting to investigate the role of the choice of boundary conditions for $\Psi(a_*)$ on the cosmological arrow of time, for DeSitter space but this issue is beyond the scope of the present paper.

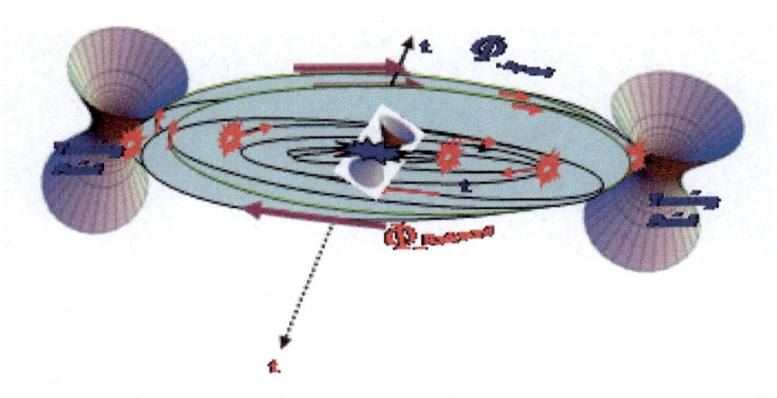

Fig. 2.2 A schematic drawing of the cycles between the two turning points of the classical path for the DeSitter wavefunction. Each cycle gets shorter due to some energy of the wavepacket being lost through the interaction with the environment. After many such cycles, the shifting of the classical path and the shortening of the cycle results in spiraling into a merging event. However, the forward and backward branches are still mixed together for each cycle. The universe always seems "aware" of its future evolution

forward and backward paths, of a universe being aware of its future evolution as it is intrinsically build onto its past. In every cycle between two turning points, the recycling universe bounces between classical and quantum in each turning point, thus a classical world would be only an intermediate stage when the universe is away from the turning point in its trajectory. Near the turning point the wavepacket would spread out by disintegrating into its many components and thus the Universe would behave as a quantum system. However decoherence in each cycle takes its toll: each cycle gets shorter and shorter due to the shifting of the energies by the interaction with states, Fig. 2.2. (The shifting of the energy is a well known result in particle physics when a particle interacts with a field then its trajectory is modified due to its energy shift by interaction [17].) It is possible in this picture that after many such cycles, with shorter and shorter trajectories, the universe will spiral in to one point and disappear. Could this speculation provide us with a mechanism for DeSitter space evaporating away its Λ during its recycling in phase space?

Acknowledgments LM-H was supported in part by DOE grant DE-FG02-06ER41418 and NSF grant PHY-0553312 and Fqxi.

2.3 Appendix

Here we summarize the calculation for the backreaction potential term $U_-(\alpha, \phi)$, of (2.3). Based on [4, 5, 14], a time parameter t is defined for WKB wavefunctions of the universe such that the equation for the perturbation modes ψ_n in the

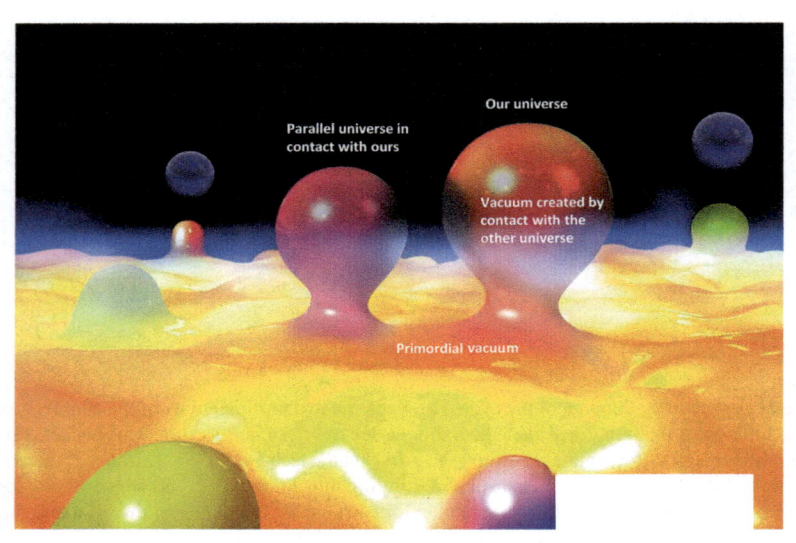

The landscape of vacua {Φ_k}. A universe is born from the wavepacket $\Psi_n[\Phi, a]$ in vacua Φ_k. The geometry of the universe is described by its scale factor $a_k[t]$.

Fig. 2.3 A schematic drawing of the birth of the universe from the multiverse. Only the high energy initial states in the multiverses are dynamically selected to give birth to a universe. Low energy states become "terminal". Notice the minisuperspace defined by the landscape vacua ϕ_k and the three-geometries $a[t]$ of universes, is the abstract space in which the wavefunctions of the universe propagate

wavefunction can be written as a Schrödinger equation and be consistent with Einstein equations.

If S is the action for the mean values α, ϕ for a nearly DeSitter state with scale factor $a = Log[\alpha]$ and vacuum energy given roughly by $\Lambda \simeq 1/2(m^2\phi^2)$, then by defining the parameter $y \equiv (\partial S/\partial \alpha)/(\partial S/\partial \phi) \sim \dot{\alpha}/\dot{\phi}$, one can write in the second order WKB approximation, [4, 5]:

$$\psi_n = e^{\frac{\alpha}{2}} \exp\left(i \frac{3}{2y} \frac{\partial S}{\partial \phi} f_n^2\right) \psi_n^{(0)}$$

$$i \frac{\partial \psi_n^{(0)}}{\partial t} = e^{-3\alpha} \left\{-\frac{1}{2} \frac{\partial^2}{\partial f_n^2} + U(\alpha, \phi) f_n^2\right\} \psi_n^{(0)}$$

$$U(\alpha, \phi) = e^{6\alpha} \left\{\left(\frac{n^2-1}{2}\right) e^{-2\alpha} + \frac{m^2}{2} + 9m^2 y^{-2}\phi^2 - 6m^2 y^{-1}\phi\right\}. \quad (2.13)$$

During the nearly DeSitter stage, $S \approx -1/3\, m e^{3\alpha} \phi_{\text{inf}} \approx -1/3 e^{3\alpha} \sqrt{\Lambda}$, where ϕ_{inf} is the value of the field during inflation, so that $y = 3\phi_{\text{inf}}$. Thus long wavelength matter fluctuations are amplified during a vacuum energy dominated universe, and driven away from their ground state [5]. This is not the case for a matter or radiation dominated universe. When the universe exits the DeSitter stage, the wavepacket for the universe is in an oscillatory regime, y is large so that the potential $U(\alpha, \phi)$ changes from $U_-(\alpha, \phi)$ to $U_+(\alpha, \phi)$, where

$$U_{\pm}(\alpha, \phi) \sim e^{6\alpha} \left[\frac{n^2 - 1}{2} e^{-2\alpha} \pm \frac{m^2}{2} \right], \qquad (2.14)$$

as can be seen from (2.13). It is straightforward to check from the equation of motion for the superHubble perturbation modes that in this case, $U = U_+$, as it was shown in [5], these modes are frozen in as in the conventional theory of perturbations. We can follow the evolution of these superHubble matter perturbation during the second stage of a vacuum energy domination phase, by using the Schrödinger equation above. During the nearly DeSitter state in the future, these modes develop gravitational instabilities because for reasons shown above, again $U(\alpha, \phi) < 0$. The matter perturbation modes f_n with $n \leq aH_0$ are coupled to the background gravitational potential, therefore the main difference with the mechanism described in [14] for the instabilities during an early DeSitter stage, is that the scale of the curvature at late times, is of the order of the low energy DeSitter Hubble horizon, $m^2 \simeq \sqrt{\Lambda}$. From (2.13) we see that during a (nearly) DeSitter stage, the patches that have $U(\alpha, \phi) < 0$, which can happen for small enough physical wave vector $k_n = ne^{-\alpha}$, develop tachyonic instabilities due to the growth of perturbations: $\psi_n \simeq e^{-\mu_n \alpha} e^{i\mu_n \phi}$, where $-\mu_n^2 = U(\alpha, \phi) f_n^2$. Thus the solution for the wavefunction of the universe, in phase space given by $\Psi(a, \phi, f_n) = \Psi_0(a, \phi) \Pi_n \psi_n(a, \phi, f_n)$ is damped in the intrinsic time α rather than oscillatory. The damping of these wavefunctions is correlated with the tachyonic, Jeans-like instabilities that develop in real spacetime for the corresponding matter perturbation modes, consistent with Einstein equations for matter perturbation f_n; when $U(\alpha, \phi) < 0$, $f_n \sim e^{\pm \mu_n t}$, while for $U(\alpha, \phi) > 0$, the long wavelength matter perturbations f_n are frozen in.

The gravitational instabilities of matter fluctuation in real spacetime, can be seen from the equation of motion for ϕ, f_n obtained by varying the action with respect to these variables. For the tachyonic case $U < 0$ universes, we have

$$\ddot{f}_n + 3H \dot{f}_n + \frac{U_-}{a^3} f_n = 0. \qquad (2.15)$$

From (2.14) we can see that only during a DeSitter stage, when $U < 0$ one obtains growing and decaying solution in spacetime roughly for $f_n \simeq e^{\pm \mu t}$ and the growth of instabilities in the infinite number of superHubble matter modes f_n with $0 < n < aH_0$ which can eventually overcome the pressure of the vacuum energy on spacetime and drive the wavefunction of the universe away from its classical trajectory.

References

1. H.V. Peiris et al., Astrophys. J. Suppl. **148**, 213 (2003) [astro-ph/0302225]
2. Sean carroll review of dark energy http://preposterousuniverse.com/reviewarticles.html
3. G.W. Gibbons, S.W. Hawking, Phys. Rev. **D15**, 2738 (1977)
4. C. Kiefer Class. Quant. Grav. **4**, 1369 (1987)
5. J.J. Halliwell, S.W. Hawking, Phys. Rev. D **31**, 1777 (1985)
6. C. Kiefer, Phys. Rev. D. **38**, 6 (1988): D.L. Wiltshire, "An introduction to quantum cosmology," Talk published in 'Cosmology:The Physics of the Universe', (World Scientific, Singapore, 1996), pp. 473–531 [arXiv:gr-qc/0101003]
7. L. Susskind, [arXiv:hep-th/03-2219]; N. Goger, M. Kleban, L. Susskind, JHEP **0307**, 056 (2003); L. Dyson, J. Lindsay, L. Susskind, JHEP **0208**, 045 (2002) [arXiv:hep-th/0202163]
8. L. Dyson, M. Kleban, L. Susskind, JHEP **0210** (2002) [arXiv:hep-th/0208013]
9. E. Witten, (2001) [arXiv: hep-th/0106109]
10. A. Albrecht, L. Sorbo, Phys. Rev. D **70**, 063528 (2004) [arXiv:hep-th/0405270]
11. C. Kiefer, H.D. Zeh, Phys. Rev. D **51**, 4145 (1995)
12. R. Bousso, JHEP **0011**, 038 (2000) [arXiv:hep-th/0010252]; R. Bousso, A. Maloney, A. Strominger, Phys. Rev. D **65**, 104039 (2002) [arXiv:hep-th/0112218]; N. Goher, M. Kleban, L. Susskind, JHEP **0307**, 056 (2003) [arXiv:hep-th/0212209]
13. N. Kaloper, M. Kleban, A. Lawrence, S. Shenker, L. Susskind, JHEP **0211**, 037 (2002) [arXiv:hep-th/0209231]
14. R. Holman, L. Mersini-Houghton [arXiv: hep-th/0511102]
15. R. Holman, L. Mersini-Houghton [arXiv: hep-th/0512070]
16. C. Kiefer, D. Polarski, A.A. Starobinsky, Phys. Rev. D **62**, 043518, (2000) [arXiv:gr-qc/9910065]
17. C. Kiefer, Phys. Rev. D **46**(4), 1658–1664 (1992)
18. B. Freivogel, L. Susskind, Phys. Rev. D **70**, 126007 (2004) [arXiv:hep-th/0408133]. R. Bousso, J. Polchinski, JHEP **0006**, 006 (2000) [arXiv:hep-th/0004134]. T. Banks, M. Dinem E. Gorbatov, JHEP **0408**, 058 (2004) [arXiv:hep-th/0309170]. M. Dine, E. Gorbatov, S. Thomas, [arXiv:hep-th/0407043]. M.R. Douglas, JHEP **0305**, 046 (2003) [arXiv:hep-th/0303194]. F. Denef, M.R. Douglas, B. Florea, JHEP **0406**, 034 (2004) [arXiv:hep-th/0404257]. F. Denef, M.R. Douglas, JHEP **0503**, 061 (2005) [arXiv:hep-th/0411183]. M.R. Douglas, "Statistical analysis of the supersymmetry breaking scale," [arXiv:hep-th/0405279]. M.R. Douglas, Comput. Rendus Phys. **5**, 965–977 (2004) [arXiv:hep-th/0409207]
19. R.M. Wald, [arXiv:gr-qc/0507094]
20. L. Susskind, "The anthropic landscape of string theory," [arXiv:hep-th/0302219]. L. Susskind, "Supersymmetry breaking in the anthropic landscape," [arXiv:hep-th/0405189]
21. R. Bousso, JHEP **9907** (1999); R. Bousso, JHEP **0104**, 035
22. S. Hannestad, L. Mersini-Houghton, Phys. Rev. **D71**, 123504 (2004) [arXiv:hep-ph/0405218]
23. M. Bastero-Gil, L. Mersini-Houghton, Phy. Rev. **D65**, 023502 (2001) [arXiv:astro-ph/0107256]; M. Bastero-Gil, K. Freese, L. Mersini-Houghton, Phys. Rev. **D68**, 123514 (2003) [arXiv:hep-ph/0306289]
24. A. Melchiorri, L. Mersini-Houghton, C.J. Odman, M. Trodden, Phys. Rev. **D68**, 043509
25. Chandra, Press Release (2006)
26. L. Mersini-Houghton, (2005) [arXiv: hep-th/0512304]
27. L. Mersini-Houghton, M. Bastero-Gil, P. Kanti, Phys. Rev. **D64**, 043508 (2001); M. Bastero-Gil, P. Frampton, L. Mersini-Houghton, Phys. Rev. **D65**, 106002 (2002); M. Bastero-Gil, L. Mersini-Houghton, Phys. Rev. **D67**, 103519 (2003)
28. R.R. Caldwell, Phys. Rev. Lett. **95**, 141301 (2005) and references therein
29. G.R. Dvali, G. Gabadadze, M. Porrati, Phys. Lett. **B485**, 208 (2000); S.M. Carroll, A. DeFelice, V. Duvvuri, D. Easson, M. Trodden, M.S. Turner, Phys. Rev. **D71**, 063513 (2005); S.M. Carroll, I. Sawicki, A. Silvestri, M. Trodden, (2006) [arXiv:astro-ph/0607458]

30. S.W. Hawking, Phys. Rev. **D32**, 2489 (1985); D.N. Page, Phys. Rev. **D32**, 2496 (1985)
31. R. Easther, E.A. Lim, M.R. Martin, JCAP **0603**, 016 (2006)
32. L. Mersini-Houghton, Class. Quant. Grav **22**, 3481 (2005) [arXiv: hep-th/0504026]; A. Kobakhidze, L. Mersini-Houghton, (2004) [arXiv:hep-th/0410213]

Chapter 3
The Future History of the Universe

Fred C. Adams

3.1 Introduction

3.1.1 Overview: Physics and Astronomy

The universe is now 13.7 billion years old. One of the remarkable achievements of modern cosmology is that we can determine this age to three significant digits. In addition to determining the cosmic age, astronomers and physicists have constructed a detailed picture of the past history of the universe, including specification of other cosmological parameters, as outlined in previous chapters. In particular, we now know the current cosmological expansion rate (the Hubble constant), the total relative density of the universe (often denoted as Ω), the temperature of cosmic background radiation, and the matter content of the cosmos, all with satisfying precision. With this current understanding of the grand sweep of astronomy, along with parallel developments in understanding the laws of physics, we can now construct a working paradigm for the future of the universe. This chapter outlines the basic features of this future timeline.

3.1.2 Copernican Time Principle

Nearly five centuries ago, Nicolaus Copernicus began a revolution that continues to this day. He developed a Heliocentric cosmology, which holds that the Sun lies at the center of the Solar System. This view eventually replaced the older picture in which our planet Earth held the central position.

F.C. Adams (✉)
Physics Department, University of Michigan, USA
e-mail: fca@umich.edu

Since the time of Copernicus, Earth's status has been continually degraded by astronomers and cosmologists. We now know that our Solar System does not lie at the center of the galaxy, but rather orbits the center at a distance of some 24,000 light years. From its position in the suburbs, our Solar System does not occupy a special place within its home galaxy.

On larger scales, this degradation continues. As discussed earlier in this book, we think that the universe as a whole is homogeneous and isotropic. In other words, the universe is "the same" everywhere in space and looks "the same" in all directions. As a result of this, our Milky Way galaxy does not occupy a special place within the cosmos. In terms of their spatial locations, our planet, our Sun, and our galaxy are aggressively ordinary.

This chapter continues this process of generalization into the temporal realm, by de-emphasizing the central position held by the current epoch. In other words, we explore the idea that the current cosmological time period has no special significance. As we shall see, the universe does indeed change markedly as it grows older. Carbon-based life has a limited – but still rather long – window for its existence. The lifetime of astronomical bodies, including stars and galaxies, is similarly restricted. Nonetheless, interesting physical processes, loosely defined as those that generate energy and entropy, could continue to operate as far into the future as we dare to imagine.

3.2 The Future of the Cosmos as a Whole

Although our universe was born within a tiny fraction of a second, a 10^{-43} second slice known as a "Planck time", it will endure for much longer. With its current age of about 14 billion years, the cosmos is already exceedingly old by the standards of the physical processes resulting in its birth. Nevertheless, chances are good that most of the life of the universe lies in its future, rather than its past, and we can now explore these possibilities.

3.2.1 Possible Future Expansion Histories

As outlined previously, the present day universe is expanding. As a result, the universe faces two basic choices for its future evolution. The future universe can either expand forever or halt its expansion and begin to collapse. Many people seem to prefer the option of eventual collapse, as it apparently allows for the possibility of another big bang event, a new beginning, and hence provides some psychological comfort. However, all the current astronomical data strongly favor the alternative scenario of continued expansion. But there is more: One of the most intriguing astronomical results of the past decade is an evidence that the universe has begun expanding at an accelerated rate. In other words, the universe is not only expanding, but also is speeding up as it does so. This discovery, which was largely unexpected, has profound consequences for the future of the universe.

Before going further, let's consider what it means for the expansion of the universe to be "accelerating". As the universe gets older, two things happen at the same time. First, the fact that the space of the universe is expanding means that any two points – think of two galaxies – are moving away from each other and hence the universe is becoming disconnected as time passes. On the other hand, as the universe grows older, light signals have more time to propagate from one place to another. Through this second effect, the universe becomes more connected with time. The universe is thus doing two things, simultaneously working to become disconnected through expansion and becoming more connected as light signals propagate. Clearly, the universe has two choices: The first option is to expand slowly, so that light signals connect points in space faster than they become disconnected due to the expansion. The second option, of course, is for the universe to expand so rapidly that the process of disconnection dominates. An accelerating universe exercises this second option.

According to our current understanding of astronomical data, our universe was expanding in its slow mode until relatively recently in cosmic history. During this period of "slow" expansion, huge volumes of the universe were connected up with light signals so that different parts of the cosmos had a chance to communicate with each other. The portion of space–time that is connected up in this manner (and that contains our galaxy) is known as the observable universe. Only the galaxies and the stars within the observable universe are close enough to have an effect – of any kind, even in principle – on experiments done here on Earth. A few billion years ago, the expansion began operating in an accelerating phase, so that distant points of the universe are now becoming less connected with increasing cosmic time. As this process continues, the implications for the universe are monumental (Fig. 3.1).

3.2.2 Large Scale Structures

If the universe continues to accelerate for times comparable to the current cosmic age, it will change its appearance in a fundamental way. At the present epoch, astronomers can observe a vast array of galaxies external to our own galaxy. Billions of moderately large galaxies, roughly comparable in size and mass to our Milky Way, can be seen across great distances. In future, however, the skies now populated with these galaxies will grow much darker.

As the universe continues its accelerated expansion, all of the galaxies that are not bound to our own will be swept outside of our cosmological horizon and will become effectively invisible. The accelerating expansion will cause the external galaxies to become disconnected from our own, so that their light signals will no longer be able to reach us. The only galaxies that will be observable in the future are those that are now gravitationally bound to the Milky Way. This bound collection of galaxies is known as the Local Group. It consists of two respectively large galaxies

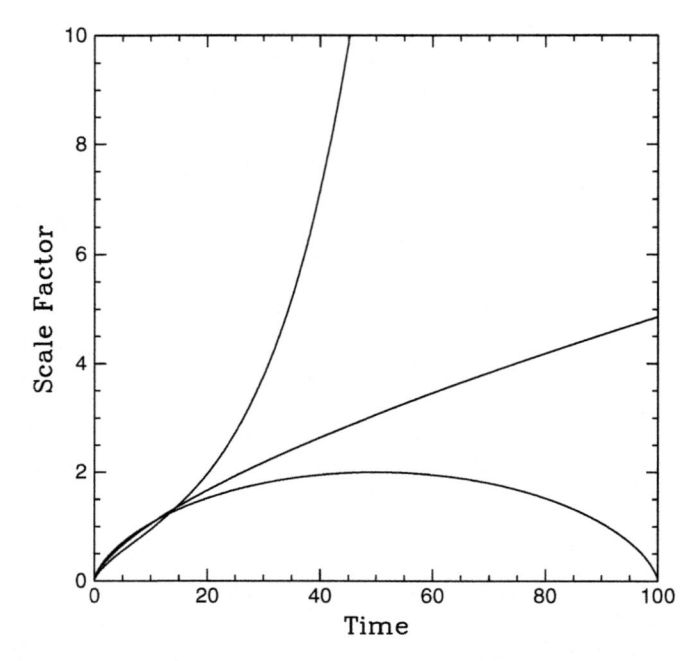

Fig. 3.1 Diagram showing three possible expansion histories for a universe. *The lower curve* shows the scale factor for a closed universe that re-collapses after a finite time. *The middle curve* shows the scale factor of a flat universe containing only matter and radiation; in this case, the universe slows down as it continues to expand and comes to a stop as time approaches infinity. *The upper curve* shows the scale factor for a universe containing Dark Energy, which causes the universe to accelerate as it expands. Current data indicate that our universe is following this upper trajectory

the Milky Way and the Andromeda, and dozens of smaller satellite dwarf galaxies. The largest and the most famous of these dwarfs are known as the Large and Small Magellanic Clouds. This impending isolation of our Local Group will take place in "only" 10 to 20 billion years, a time scale roughly comparable to the current cosmic age.

This separation of galaxy groups will occur across the entire universe, so that each cluster of galaxies will become isolated. Not only will these clusters not be able to communicate with each other but also even in principle, will not be able to see each other. At least as long as the universe continues to expand in its present mode of acceleration. Each bound structure in the universe today is thus destined to become its own island universe in the future. Our local group is rather a modest-sized cluster of galaxies, with a total mass of trillions of Suns. The largest bound structures in our universe today, which will become the most populated island universes of the future, are almost a thousand times more massive. These structures are the largest ones that the universe can ever produce (Fig. 3.2).

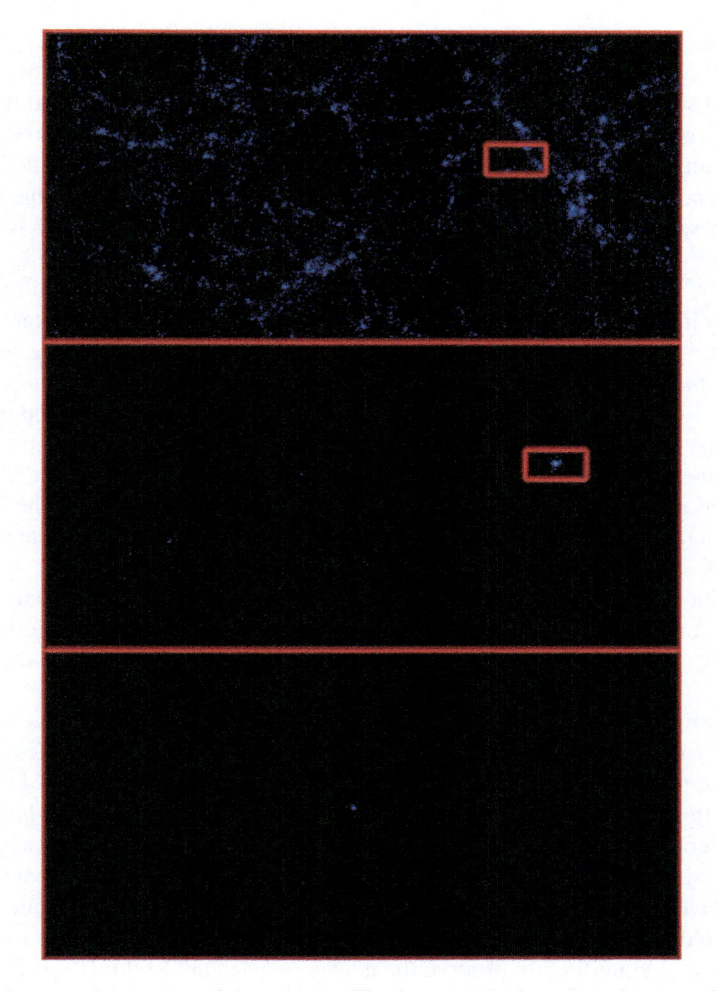

Fig. 3.2 Large scale structure of the universe. The three panels show the universe at the present epoch (*top*), in the near future (*middle*), and at a future cosmic age of 100 billion years (roughly ten times the current age). Each panel shows an expanded view of a portion of the previous panel, as indicated by the red frames

3.2.3 Return to a Steady State Universe

As outlined in previous chapters, the story of our universe is described by what is known as the Big Bang Theory. The experimental support for this paradigm rests on three main pillars. The first pillar of evidence is that the universe is observed to be expanding. The expansion of the universe was discovered in the 1920s by Edwin Hubble, Vesto Slipher, and others, and has been continually re-measured to the present day.

Because the cosmos is large today, and growing larger, one can conceptually run the clock backwards and see that the universe of the past was smaller, denser, and hotter. At sufficiently high densities, radiation dominates over matter and hence the early universe must reside in a state of thermal equilibrium. If the universe went through an early hot and dense phase, the cosmos should now be filled with a background sea of radiation left over from its early history. This cosmic background radiation was discovered in the 1960s by Arno Penzias and Robert Wilson, and provides a compelling piece of evidence in favor of the Big Bang.

Finally, our present day universe is observed to contain a great deal of helium and other light elements. Helium can be synthesized in stars – after all, stars operate by "burning" hydrogen into helium. However, the amount of helium we see in the universe today is far more than that produced by all the stars, in all the galaxies, over all of cosmic time. At least so far. The only way to account for the observed abundance of helium is that it was produced early in the history of the universe. During the first three minutes of cosmic evolution, the universe was hot and dense enough to sustain thermonuclear reactions. In addition, the amount of helium that is predicted to by synthesized during this early nuclear epoch is consistent with that observed.

Resting on the three experimental pillars outlined above, the Big Bang Theory provides a remarkably successful description of the past history of the universe. In the future, however, all these points of confirmation will be far more difficult to observe.

As outlined above, due to the accelerating expansion of the universe, our local group of galaxies will soon become its own island universe. In this context, "soon" is relative to the vast sweep of time to come, and here refers to tens of billions of years. After the external galaxies are swept outside our cosmological horizon by the unrelenting and accelerating cosmic growth, there will no longer be any visible beacons by which to measure the expansion. Today we deduce the expansion rate of the universe by measuring how fast external galaxies are receding from our own galaxy. In the future, with no external galaxies to see, any observers that are present will not be able to directly observe the cosmic expansion. In this manner, the first pillar of Big Bang theory is compromised in our cosmic future.

The second pillar of evidence for the Big Bang, the cosmic background radiation, will also be much harder to detect in the future. As the universe expands, the wavelength of this background radiation is stretched along with the background of space. Right now, the cosmic radiation has wavelengths of about one centimeter. This frequency of this radiation is about ten thousand times too small for human eyes to see, but can be readily detected with modern technology. If you wait long enough, the wavelengths of the cosmic background radiation will be stretched to become larger than the galaxy. And then larger yet. With such enormous sizes, the photons (particles of light) will become exceedingly difficult to detect with any instrument. The background radiation fields will thus be redshifted into oblivion, and another piece of evidence in favor of the Big Bang will become obscure.

Our universe today is not old enough for stars to have produced the abundance of helium we observe. But we can understand its presence from the hot early days of the past. This early three minute window of opportunity is known as Big Bang Nucleosynthesis. As the cosmos ages, and stars produce greater amounts of nuclear waste, the helium produced by the Big Bang and that produced by stars will become increasingly difficult to separate. In rough terms, the early phase of Big Bang Nucleosynthesis processed one fourth of the available hydrogen into helium (where the accounting is done by mass, rather than by number of particles). To date, stars have burned only a few percent of the total store of hydrogen into helium, so that the early era of production dominates the supply. After the universe is old enough and stars have produced most of the helium, however, it will become increasingly difficult for future astrophysicists to deduce the effects of the early universe and thereby to determine the past history of the cosmos. In this manner, Big Bang Nucleosynthesis will also be compromised.

On the whole, the above changes will not only remove – or at least veil – three of the main pieces of evidence for the Big Bang, but also make the universe of the future appear much like the steady state model. The background universe will seem sedate and unchanging when viewed from our Local Group. The universe will continue to expand, however, it will just be more difficult to determine that it is doing so.

3.2.4 Future Phase Transitions

Although current observational data indicates that the expansion of the universe is accelerating, and the reason for the acceleration is far from certain. Ordinary matter, through its gravitational attraction, causes the expansion to slow down with time. In order to make the expansion speed up, space must be filled with a counter intuitive type of energy. This substance is generally called "Dark Energy", although our understanding of its nature remains vague. Roughly speaking, Dark Energy is the energy associated with empty space, which of course means that empty space is not really empty; instead it is filled with *something*, and that something has a negative pressure. In general relativity, both mass and pressure have gravitational effects. The fact that this energy causes the universe to accelerate is thus due to the odd properties of gravity for a substance with a negative pressure – and we call this bizarre substance Dark Energy.

The idea of empty space having an associated energy arises in other contexts. Quantum mechanics tells us that virtual particles pop in and out of existence in empty space. Further, these virtual particles, or more specifically their effects, can be observed in particle accelerator experiments. In this sense, virtual particles are real. Generally speaking the net result of these virtual particles can add up, to produce a non-zero energy level for the vacuum. This energy is the so-called Dark Energy.

Once we allow for the vacuum – empty space – to have an energy level, it is not much of a stretch to imagine that it has two or more accessible energy levels. Suppose empty space has two energy levels, where the universe is now in a state corresponding to the higher level. If the universe, or more precisely the vacuum state of the universe, is in the high energy state, then it is possible for it to make a transition to the lower energy state. The result would be a phase transition, where the energy associated with empty space would change across the entire universe. In such an event, the transition would first occur within small bubbles of space–time, where empty space has the new, low energy state within the bubble, and maintains the old, higher energy level outside. Under the right conditions, the bubbles can grow with time, and eventually percolate, thereby converting the background space of the universe entirely into the new, low energy state. The phase transition of the vacuum is thus like ice crystals nucleating within a sea of super-cooled water. The icy crystals are small at first, but eventually grow and take over the entire volume of water. In the case of the cosmos, when the phase transition runs to completion, the vacuum attains a new energy state, which has potentially interesting consequences.

These consequences are speculative, but far-reaching. One possibility is that the phase transition could induce changes in the laws of physics. Although this possibility is difficult to imagine, one can think about it as follows. The energy levels of empty space, the same energy levels that change during the phase transition, can be described as a field – think of this field an analog of the magnetic fields that are produced by the Earth and other sources (although there are some differences). If the field is coupled with any parameters that affect the laws of physics, then the laws will change as the phase transitions takes place because the effective value of the field will change. Possible examples include changes in the strengths of the fundamental forces, or changes in the masses of the fundamental particles. At the present time, physicists are still working to understand the energy level or levels associated with empty space, and a full working theory of this phenomenon is not available. As a result, we simply do not know whether or not such a phase transition is possible, or if it will result in variations in the laws of physics (Fig. 3.3).

3.2.5 Heat Death

In the nineteenth century, before the development of Big Bang theory, the "heat death" of the universe was one of the problematic issues facing physical science. This issue can be encapsulated by the following question: Given that all physical systems tend to evolve towards thermal equilibrium, why hasn't our universe done so? In other words, why isn't our universe in thermal equilibrium, with all bodies at the same temperature? Before resolving this paradox, let's examine its parts in greater detail. The tendency toward equilibrium should be familiar: Ice cubes in a glass of water will melt, thereby cooling the water and heating the now-liquid ice, until the whole system reaches the same temperature. In an analogous manner, the universe should adjust itself to reach a uniform temperature.

Why does this matter? If different parts of the universe have different temperatures, then a "heat engine" can be run between the two regions. In rough terms, again, the two parts of the universe will transfer heat in order to reach a uniform temperature, and in doing so, some useful work can be extracted. A universe with

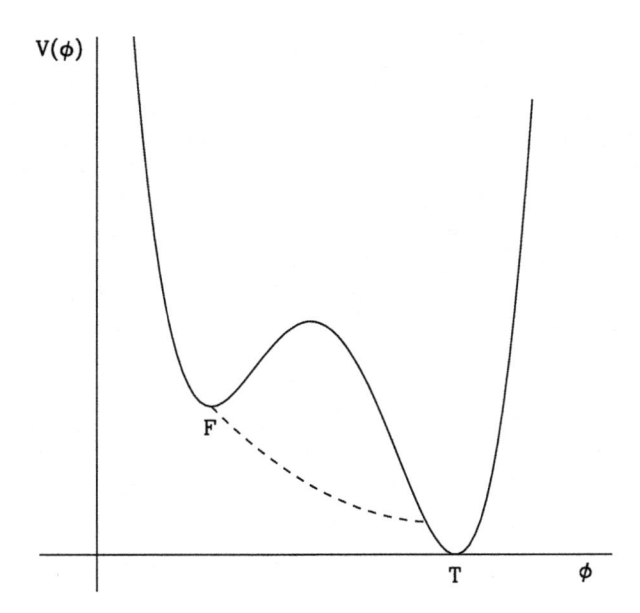

Fig. 3.3 The potential energy of the vacuum state of the universe during a cosmic phase transition. This plot shows the potential energy $V(\phi)$ as a function of the scalar field ϕ that determined the vacuum energy density of the universe. In this (hypothetical) case, the universe currently lives in the high energy state marked by the letter F (which indicates the false vacuum). In the future, the universe can make a transition to the lower energy state marked by the letter T (which indicates the true – or lowest energy – vacuum state)

different temperatures thus has the potential to do useful work and can be an interesting place. Fortunately, our present-day universe is such a place: Radiation pours out of the surfaces of hot stars, with photospheric temperatures of thousands of degrees Kelvin, while the background universe has a much lower temperature of only about 3 Kelvin. Our universe is thus far from equilibrium, and can do useful work, such as drive biological evolution on habitable planets like our Earth. If the universe were to reach thermal equilibrium, however, all such processes would shut down, and the cosmos would become markedly less exciting.

Before the advent of Big Bang theory, many scientists thought that the universe was infinitely old, which provides a long time for it to reach thermal equilibrium. Hence the paradox. In the context of an expanding universe, however, the concept of heat death changes. The first modification is that the universe has a finite age, about 14 billion years, so that it simply has not had time to reach thermal equilibrium. In addition, since the universe is expanding, the background temperature of the universe – that provided by the cosmic background radiation – is continually becoming colder, which provides new opportunities for heat flow. However, it remains possible for the universe to reach a state in which it expands adiabatically, which means that no new entropy is produced and the universe would (again) become a boring place. This possibility is sometimes called a "cosmological heat death", but our present day universe is far from such a state.

To complicate the discussion, our universe is now thought to be dominated by dark vacuum energy, which provides a new type of radiation. If the energy density of the vacuum is constant, or nearly so, it can produce photons of extremely long wavelength that provide a background sea of radiation. This energy is emitted through a process that is analogous to the Hawking effect. This mechanism allows black holes to slowly lose their mass (see below). In this case, however, the wavelength of the emitted radiation is comparable to the size of the universe of today. The cosmic background radiation, left over from the hot early phases of the big bang, completely dominates this second type of radiation produced by the vacuum. In the far future, when the universe is about one quadrillion times larger than it is today, the radiation emitted by the vacuum will provide a reservoir of constant background temperature to the universe. After this milestone is reached, the universe has a chance to reach thermal equilibrium and thereby experience heat death, but not until a large number of interesting physical processes (and perhaps biological processes) have run their course.

3.2.6 The Big Rip?

Although the phase transitions described above are speculative, even more exotic possibilities exist. One of the most dramatic scenarios is based on the idea that the vacuum energy level need not be constant in time. In most cases, one would expect the energy of empty space to either remain constant or decrease as the universe grows older. However, this need not be the case. Given our current physical understanding of these issues, it remains possible – but not necessarily likely – that the energy level of empty space will actually *increase* with time.

If faced with increasing levels of vacuum energy, the universe would experience a far different future. In any expanding universe, including our own, a constant battle is waged between gravity, which acts to bind cosmic structures together, and the expansion itself, which acts to pull structures apart. The organizational forces of gravity win this celestial war only when the density of the cosmic structure is sufficiently high. As one example, the galactic halos in our present-day universe are dense enough to resist the cosmic expansion and remain intact. Within the galaxies, the densities are even higher, and the current levels of vacuum energy density have essentially no effect. Outside the galaxies, however, the densities are low, and the expanding background of the universe stretches the galaxies farther apart from each other.

If the vacuum energy density increases with time, the expansion rate also increases with time. Any structure with a fixed density will eventually fall behind, so that the growing density of the vacuum will ultimately become larger than that of any given astronomical object. First galactic clusters, and then galaxies, will be ripped apart by the ever-increasing severity of cosmic expansion. Although stars are denser than the galaxies by an almost unfathomably large factor of 10^{24}, if the vacuum energy levels continue to increase, the stars themselves will be ripped apart. This destruction continues as long as the increase in vacuum energy continues. Eventually, even atoms and then atomic nuclei would be ripped apart in this scenario, leaving no structure of any kind.

This particular scenario for cosmic doomsday is possible, and intriguing, but this is not the most likely fate for our universe. For the remainder of this discussion, we thus consider the case in which the energy of empty space does *not* grow without bound. In this event, the cosmos is destined to expand forever, or at least for an enormously long time. The expansion rate may even accelerate, as it is doing at the present cosmological epoch, but the galaxies, stars, and planets will endure. We can then ask what their ultimate fate will be? These questions are addressed in the subsequent discussion.

3.3 The Future of Galaxies

Galaxies are not expanding with the background of the universe. Instead, they have won their battle against the cosmic expansion and have separated themselves into their own little islands. Well, not quite: Galaxies tend to live in somewhat larger structures, clusters containing many individual galaxies, but these clusters have indeed separated themselves from the background universe. Within these gravitationally bound entities, a rich future is waiting to unfold.

3.3.1 Cosmic Isolation

The universe today is expanding and the rate of the expansion is increasing with time. In such a universe, all of cosmic structure is being pulled apart. On sufficiently small spatial scales, the densities are large enough that gravity can fight this accelerating expansion and hold things together. As a result, planets, stars, and galaxies are not in danger of being destroyed by the expansion taking place in the background universe. On somewhat larger scales things can be different.

The largest gravitational bound structures in our universe today are the galaxy clusters, large entities containing perhaps hundreds of galaxies, huge amounts of hot gas between the galaxies, and a corresponding complement of Dark Matter. Although these structures are robust enough to withstand the cosmic expansion, they will not remain bound to each other. Instead, each galaxy cluster will be accelerated away from every other galaxy cluster. In the end, each of these gravitationally bound structures will become isolated and thereby forming its own "island universe".

In this context, when galaxy clusters become isolated, they not only are separated by vast distances, but also accelerate away from each other at enormous rates. In fact, the expansion rate is rapid enough that these distant clusters move away from each faster than light signals can propagate between them. In other words, galaxy clusters of the future will not be able to communicate with each other, even in principle, even over an infinite span of future time. Moreover, this isolation will happen in the relatively "near" future, when the universe is only a few times older than its present age of 14 billion years.

This claim of isolation seems to defy common sense. How can clusters move away from each other so quickly that light cannot travel between them? Doesn't Einstein's theory of relativity tell us that nothing can travel faster than the speed of light? Yes, it remains true that no information can travel faster than the speed of light. In this case, however, the galaxy clusters are not really moving. They are fixed at their particular locations in space. In rough terms, the expanding universe around them can be considered to be creating new space in between the existing clusters – or, equivalently, stretching out the space between them. In any case, the clusters are becoming farther apart, but they are not moving through space at all. As a result, their rapid recession from each other does not violate the cosmic speed limit. But still the clusters grow farther apart, and they do so at such a rapid rate that light signals cannot connect them up. They become isolated structures, effectively cut-off from communication with the rest of the universe.

3.3.2 Collisions Within Clusters

Although clusters of galaxies will become isolated from each other, the galaxies within clusters will continue to interact. In fact, as they participate in a complicated series of orbits, the galaxies sometimes collide with each other. Over time, these collisions reshape the galaxies inside every bound cluster. Within our Local Group of galaxies – the cluster that contains our Milky Way galaxy and our solar system – one such collision is imminent (Fig. 3.4).

Within the Local Group, the two principal actors are our Milky Way and its sister galaxy, Andromeda. These two galaxies are gravitationally bound to each other and orbit within the collective gravitational potential of the Local Group. In many ways, Andromeda is much like our Milky Way. It has a similar mass, and size, and shape. In addition, Andromeda has one particularly interesting characteristic – it is coming straight toward us! The Milky Way and Andromeda are scheduled to collide within the next Hubble time or two, where a "Hubble time" is comparable to the current age of the universe. Given the uncertainties in the measured velocity of Andromeda, it could collide with the Milky Way on its first pass, or it could miss. Given that the two galaxies are bound to each other, however, it would then collide with the Milky Way on the next pass. In the long term, the two galaxies are destined to collide and merge.

What happens when two galaxies collide? When viewed from the outside, the result of a galaxy collisions looks like a train wreck. The beautiful, well-ordered spiral patterns are destroyed, and long streamers of stars and gas are ripped out of the galactic disks. After a few orbits, the colliding galaxies merge into an amorphous pile of stars. The gas clouds within the galaxies also tend to collide, with the result being a burst of star formation. After the collision, however, relatively little gas

Fig. 3.4 Galaxy collision. This picture shows two galaxies is in the process of colliding. A similar collision is scheduled to take place in the future between our Milky Way and its sister galaxy Andromeda

remains within the merged galaxy itself, and star formation activity grows markedly lower. With few newly born stars, the merger product becomes steadily redder as its stellar population ages.

When viewed from within the galaxy, however, the collision is much less dramatic. One must keep in mind that galaxies are almost relentlessly empty. If you were to reduce the size of stars down to the size of sand grains, the typical distance between stars would be several miles. Imagine your friend standing a few miles away, and holding a sand grain, while you try to throw a sand grain of your own and hit it. The chances that you would succeed are rather low. In much the same way, chances are low that individual stars will actually run into each other when two galaxies collide. Instead, the main result is a gradual brightening of the night sky. Even this effect is less than dramatic: The collision takes place over millions of years, and the brightness increases by only about a factor of two.

3.3.3 The End of Star Formation

At the present cosmological epoch, stars are being formed within dense clouds of gas in our galaxy, and others. The galactic disk contains a healthy supply of gas, with its usual cosmic abundance of hydrogen, helium, and small amounts of heavier elements. The gas cycles through various phases in the interstellar medium, and the material between the stars. Some of the gas is hot and diffuse, with temperatures of a million degrees. Other gas is cooler and somewhat denser. At sufficiently high densities and cold temperatures, the hydrogen gas forms molecules, and the resulting clouds of gas are called molecular clouds. These clouds, typically with the mass of a million Suns, are the sites of ongoing star formation in the galaxy.

The mass supply of the galaxy cycles through the stellar population. As more stars are manufactured, a great deal of material is locked up within them. However, the stars return some fraction of their mass back to the interstellar medium. During their early formative stages, while they are still assembling their masses, stars return about a third of their mass back into their parental clouds in the form of winds and outflows. During the bulk of their lives as hydrogen burning furnaces, these winds and their accompanying mass loss is much more modest. At the ends of their lives, however, stars ramp up their activity once again, and return a significant fraction of their mass back to the galaxy. The amount of material involved depends on the stellar mass. The smallest stars retain most of their mass. Larger stars not only return more mass, but also a larger fraction of their mass. For example, a star with an initial mass of eight Suns loses about 80% of its material after it ceases burning hydrogen into helium in its central core. The mass lost by the stars, during all phases of their lives, is returned to the interstellar medium for future use.

As the ecological cycle of stellar evolution proceeds, an ever increasing fraction of the mass supply becomes locked up in small stars, in the stellar remnants left over after most stars die, and in brown dwarfs. These latter objects are too small to sustain hydrogen fusion in their cores, so they just sit around and take up space. Since hydrogen gas is the raw material needed to make new stars, and the gas supply decreases with time, this karmic cycle must eventually wind to a close.

Given the present day supply of unburned hydrogen, the galaxy would run out of gas – literally – on a time comparable to the current age of the universe if stars continued forming at the present rate. Although the galactic disk will gain some additional gas supplies in the future, due to material falling into the galaxy from its halo, these meager additions will not be sufficient to stave-off the upcoming energy crisis. A more important effect is that galaxies tend to practice conservation of their natural resources: As the supplies of gas decrease, the rate of the star formation also decreases, so it will take a rather long time for our Milky Way to make its last star. This end must come, however, and optimistic estimates indicate that after several trillion years, ordinary star formation will no longer be operational in our galaxy, and others across the universe.

3.3.4 Power Output of Galaxies

Over the next hundred billion years the power output of the galaxy will remain remarkably constant. At first glance, this finding might seem counter-intuitive. The rate of star formation will decrease with time, so that dying stars are not readily replaced. The largest stars provide the most power in the universe today, but they live for a much shorter time than the smaller stars. The largest, more luminous stars live for only a few million years. As the galaxy ages, the fraction of high mass stars decreases and the stellar population shifts towards stars of lower mass. And these smaller stars are dimmer than their larger brethren.

So why doesn't the total radiation output of the galaxy fall quickly? Even today, the low mass stars are the most numerous, and they will increase their domination in the future. Because the smallest stars live for much longer than the current age of the universe, the galaxy retains their services for trillions of years. To put this span of time into perspective, these diminutive stellar objects will sustain nuclear reactions in their cores for more than one thousand times longer than the universe has been in existence. In addition to their population increase, these stars get brighter with time. This behavior is shared by all stars, but the smallest stars slowly increase their power output over trillions of years. As a result as the larger stars grow old and die away, the smaller stars grow brighter and increase their numbers somewhat. The increase in power from small stars nearly compensates for the decrease from larger stars, so that the total radiative output of the galaxy is approximately constant until the low mass stars go off-line.

Note that the lifetime of the longest-lived stars is roughly comparable to the time over which the galaxy can continue to make new stars. After the galaxy becomes tens of trillions of years old, it runs out of gas to make new stars and its longest-lived stars fade away. After this time, the brightness of the galaxy drops precipitously. This transition provides an important milestone. When the universe is younger than a few trillion years, the stars will be shining brightly. Afterwards, nuclear burning stars will be gone, star formation will shut down, and the character of the universe will change accordingly.

3.3.5 Evaporation of Galactic Disks

In the longer term, galaxies themselves are scheduled to evaporate into oblivion. As the stars within galaxies trace through their orbits, they pass by other stars, and the stellar bodies pull on each other with their gravitational fields. One can think of this process as a relatively gentle scattering process, where the stars scatter-off each other from afar. While any one of these events has little consequence, their effects build up over the vast spans of time available in the future. Through this cosmic dance, the stars share their orbital energy with each other. Through this distribution of the energetic wealth, some stars are promoted to high energies, high enough to leave the galaxy. The stars that have lower energies are left behind. Due to the nature of gravitational forces – specifically, the property that self-gravitating systems have a negative specific heat – the "cold" stars that are left behind will continue to scatter-off each other and the system will continue to eject stars. In this manner, the galaxy literally evaporates away with time. After about 10^{20} or perhaps 10^{21} years, the bulk of the galaxy will be evaporated. Across the universe, any remaining galactic disks will become unrecognizable.

3.3.6 Demise of Galactic Halos

The evaporation of galactic disks does not represent the final chapter in the lives of galaxies. The bulk of the matter in galaxies is not contained in stars, but rather in weakly interacting Dark Matter particles. These stealthy particles make up Dark Matter halos, large structures that extend far beyond the visible portions of galaxies marked by stars, gas, and other forms of ordinary matter. These Dark Matter halos live somewhat longer than the time required for galactic disks to evaporate. In the end, however, they too will be gone.

The ultimate end of Dark Matter takes place because the particles can annihilate with each other. Two Dark Matter particles can interact when they pass sufficiently close to each other, and such interactions can lead to total annihilation, i.e., the conversion of all their mass into small bursts of radiation. These interactions are rare. Dark Matter particles are thought to interact only through gravity and the weak force, where the latter drives their destruction. Since the weak force is well, weak, only acting over a short distance, and since the densities are extremely low, it takes an enormously long time for Dark Matter particles in the halo to be destroyed.

In the halo, the Dark Matter density is less than one particle per cubic centimeter and the typical speeds are about 200 km/s. With these parameters, and the strength of the weak force, it takes about 10^{23} years for Dark Matter annihilation to change the nature of galactic halos. When Dark Matter particles annihilate, their mass energy is converted into radiation, which in turn streams away. The galaxy is left with less mass and hence less gravitational pull. This slimmed down version of the galaxy expands somewhat and slowly loses its stars to the background universe.

In addition to direct destruction, Dark Matter evolution takes place through an indirect channel. In this latter process, Dark Matter particles are captured by dead stellar remnants, where they sink to the central core and subsequently experience annihilation. This process provides the stellar remnants – and their galaxies – with an additional energy source that is discussed in the following section.

3.4 The Future of Stars

Having looked into the future at the larger scales of the cosmos, we now focus our attention closer to home. Here we find that stars are the fundamental building blocks of the galaxies and indeed the universe. At the present cosmological epoch, stars provide most of the energy that is being generated. Even when we gaze across the universe to view distant galaxies, the starlight from those galaxies is often the main attraction. The present-day universe is very much dominated by stars, and these objects will continue to run the show for some time into our cosmic future.

Fig. 3.5 Red giant sun. This figure depicts an artist's conception (from B. Jacobs) of the future of our planet with the Sun becoming a red giant in the background. The Sun will swell into this configuration about 7 billion years from now. At this future time, the biosphere will have long time from being sterilized

3.4.1 The Red Giant Sun

Perhaps the most immediate concern is the fate of our Sun, our planet, and the rest of the Solar System. In the present time, the Sun provides power for our biosphere, and Earth is a pleasant, habitable planet. In the relatively "near" future, however, this favorable set of affairs is slated to change. In about 3 billion years, the Sun will become about 40% brighter than at present. This change is not surprising, as all stars grow brighter as they age. This is what stars do. However, this seemingly modest difference in brightness has profound consequences, as it marks the end of life on Earth! With this increase in power, the Sun will drive a runaway greenhouse effect on Earth. The oceans will boil and the entire biosphere will be efficiently sterilized (Fig. 3.5).

Although this event marks the end of life on Earth, the planet itself will survive relatively unscathed. The Earth, considered as a planet, does not particularly care whether living organisms grace its surface or not. It will live comfortably through this era and will last for another four billion years. But the Earth is not destined to live forever.

The event that endangers the survival of Earth itself takes place when the Sun begins to exhaust its supply of hydrogen fuel in its central core. At this epoch, the Sun will begin its journey towards becoming a red giant star. As the fuel supply in its central furnace becomes depleted, the star cannot maintain enough nuclear reactions to provide sufficient pressure to hold up its immense weight. The central regions of the star will then shrink, while the outer layers expand. The growing Sun will first extend out to the radius of Mercury's orbit, and our innermost planet will be swallowed by the inflating Sun. A short time later, the same fate awaits Venus.

The red giant Sun will continue its expansion until its outer surface is as large as the current radius of Earth's orbit. But Earth will no longer be there. As the Sun expands, it loses mass, which holds the planets in their current orbital positions. With a lighter Sun, Earth tends to slip out to an orbit of larger radius. At the same time, however, the mass that flows-off of the Solar surface provides a headwind for Earth to plow through. This interaction leads to a frictional drag acting on our planet, and thus tends to move the Earth to an orbit with lower energy, in other words, closer to the Sun. Although the interplay between these two effects is not completely known, odds are good that Earth will be dragged back into the solar photosphere at this future time. After 10 billion years of history, including the rise of mankind, Earth will be gone. The final result of our planet's existence will be to add its metal content to that of the Sun.

Although this dismal scenario will not play itself out for many billions of years, some might find it disconcerting to contemplate the end of our planet. However, this fate, although likely, is not entirely sealed. There remains a chance that Earth can be saved.

3.4.2 Planet Scattering

Although the galaxy is incredibly empty, over vast expanses of future time, stars in the solar neighborhood can make close passages. If passing stars wander close enough to the Sun, and if the phases of the planets in their orbits are in the right configuration, then Earth's orbit can be seriously disrupted. In the extreme case, Earth can be ejected from the Solar System altogether. The chances of this event taking place are rather low: During the 3 billion year window that life on Earth has remaining, the odds of Earth being ejected by passing stars are about 1 part in one hundred thousand (10^5). Although these odds are low, one should keep in mind that they are still much greater than the chances of winning a typical state-wide lottery.

From the point of view of the planet, leaving the Solar System is clearly a saving event. An ejected Earth will not be swallowed during the red giant phase of the Sun, and the planet will continue its existence. From the point of view of life on the Earth's surface, however, the prospects are not so rosy. In approximately one million years, the oceans of the ejected Earth will freeze and the biosphere will be almost entirely gone. Deep inside the Earth, however, natural radioactivity provides an internal heat source. With our planet's favorable position at the present time, the Sun provides about ten thousand times more power than the Earth's interior, so the latter has little effect on surface life. If Earth leaves the Solar System, however, the internal radioactivity will be the only power source left, and it can keep small pockets of water in liquid form far below the Earth's surface. It even remains possible for such pools of water to support life, albeit only at the microbial level. In fact, life on Earth might even continue over a longer time span, if Earth is ejected from the Solar System, compared to the 3 billion years remaining, if Earth stays in its current orbit.

Although Earth ejection is a low probability event, and it leads to a saving of the planet, the real big-money pay-off comes from a passing star capturing the Earth. Over the same 3 billion year time period while our biosphere remains viable, the odds of a capture event are about one part in three million, roughly comparable to the odds of winning the lottery. Of course, the odds of being captured into a habitable orbit are considerably lower.

Planet scattering tells us much about the dynamics of solar systems in the face of perturbations from other stars. One thing we learn is that the most important channel for Earth ejection is not through direct disruption, where the passing star pulls on Earth itself and thereby changes its orbit. Instead, it is much more likely for a passing star to influence Jupiter, because it resides in a larger orbit and presents a larger effective target area. However, Jupiter is both close enough (to Earth) and large enough (in mass) that changes in the Jovian orbit can easily lead to ejection of Earth from its orbit. In this event, which is the most likely scenario for ejecting Earth, Jupiter acts as an intermediary between the passing star and our planet.

Why are such dynamical scenarios of interest? Planet scattering of the type considered here can play a role in determining the orbits of planetary systems across the galaxy. Most of the galaxy is sparsely populated, with a stellar density of about one star per cubic light year. However, a sizable fraction of stars and their solar systems are born within star clusters, where they reside for their first 5–10 million years of life. In such clusters, the stellar density, and the chances for close passages like those described above, are much higher. In spite of these increased odds, the smaller time window makes such effects relatively modest. Nonetheless, they do in fact take place. Sculpting solar systems by passing stars in their birth clusters is one of many processes that contribute to the rich diversity of solar system architectures that are now being discovered.

3.4.3 *Future Evolution of Red Dwarfs*

Most people have the impression that our Sun is an average star. While this statement is true, in some sense, it is also misleading: In the collection of the nearest 50 stars, our Sun is the fourth largest by mass. By virtue of this ranking, the Sun is actually a respectably *large* star! As a result, the vast majority of the stars in the sky are smaller than the Sun. Not only that, smaller stars live longer than large stars, and the smallest stars live much longer. Stars belonging to this smallest class are known as "M stars" or "red dwarfs" and they will inherit the universe in the relatively near future.

The hydrogen burning in the lifetime of a star is given by a simple ratio: The stellar mass divided by the stellar power. The star's mass determines how much hydrogen fuel it has to burn, through nuclear fusion, and the power determines the rate at which it burns. With the proper conversion factors, this ratio thus determines how long stars can live as viable nuclear furnaces. For example, our Sun is scheduled to burn for a total lifetime of 10–12 billion years. This time is more than twice the current 4.6 billion year age of the solar system and roughly comparable to the current age of the universe.

One key feature of stellar populations is that the smallest stars are not only dimmer than the larger stars, they are much dimmer. Across the mass range for stars, from ten times smaller (in mass) than the Sun to one hundred times larger, the power output of the star is roughly proportional to the cube of the mass. Typical red dwarf stars are thus one thousand times dimmer than the Sun. If they had the same fuel supply, these small stars would live one thousand times longer than the total lifespan of the Sun, or equivalently, one thousand times the current age of the universe.

One of the surprises resulting from studies of future stellar evolution is that these small stars *do* have approximately the same fuel supply as our Sun. In stars like the Sun, only the central core region, about 10% of the mass, is hot enough to sustain nuclear reactions. These stars are also stable to convection, so that unprocessed nuclear fuel is not delivered to the core from the layers above. As a result, the Sun has access to only 10% of its hydrogen fuel supply. Smaller stars, about ten times smaller than the Sun, remain convective for most of the lives. They continuously cycle nuclear fuel from all radial levels down into their stellar cores. Red dwarfs thus have access to nearly all of their hydrogen supply, making their nuclear arsenal comparable to that of solar-type stars.

As a result, the smallest, longest-lived stars will burn for about one thousand times the current age of the universe. As mentioned in the previous section, stars also get brighter as they age, so that the total power output of the galaxy will stay roughly constant over this span of time. Furthermore, the longest time for which galaxies can sustain ordinary star formation is comparable to the lifetime of the red dwarf stars. When the universe is younger than about ten trillion years, galaxies can make new stars and the longest-lived stars are still around. When the universe grows older than this age, star formation and stellar evolution grind to a halt. The universe then becomes a darker, less energetic place.

3.4.4 Inventory of Degenerate Objects

After the universe becomes many trillions of years old, conventional star formation will have come to an end, and all existing stars will have exhausted their supplies of hydrogen fuel. This time in the future history of the universe thus marks the end of an era, the Stelliferous Era, when stars provide the most important energy source for the cosmos. After the stars burn out, their role is filled by the stellar remnants they leave behind (Fig. 3.6).

Such stellar remnants are called degenerate objects. This term does not make a moral statement about this future epoch of the universe. Instead, it refers to

3 The Future History of the Universe   91

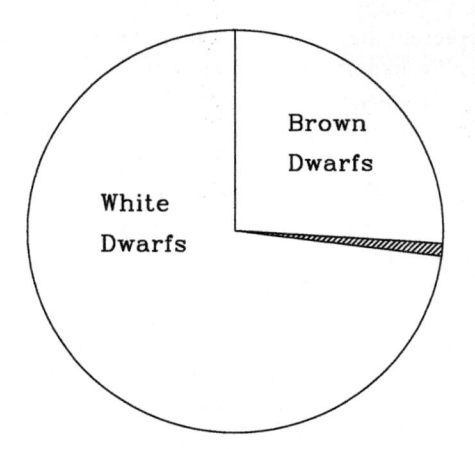

Fig. 3.6 Pie chart showing the inventory of degenerate stellar remnants – the objects left over after stellar evolution has run its course. The population of objects includes white dwarfs (largest sector), brown dwarfs (next largest sector), as well as neutron stars and black holes (depicted by the narrow shaded sector). This inventory applies at the start of the Degenerate Era

the source of internal pressure that holds up these objects. After a star dies, the remaining stellar corpse condenses until its density is about one million times denser than water. In other words, an object with a mass comparable to the Sun is squeezed into a volume of about the size of the Earth. At this enormous density, the electrons in the outer shells of atoms are so close together that their quantum mechanical properties become important. As a result, the electrons exhibit wave-like properties, and one result of this behavior is to produce a pressure force that holds up the star. This process of producing pressure via quantum effects is known as "quantum mechanical degeneracy pressure", and it provides the internal support for most of the objects that populate this future era.

After the stars burn out, the universe still contains four main types of stellar bodies. None of them are true stars, in that they cannot sustain nuclear reactions, but they are roughly the mass of today's stars and are the principal actors during this future time. White dwarfs are the first type of degenerate object and they are the most numerous, both by number and especially by their mass contribution. Approximately 997 out of every 1,000 stars will turn into white dwarfs upon their death. These stellar remnants typically retain somewhat less mass than that of our Sun, but they are much smaller in radius and are one million times denser. The mass range for white dwarfs is quite compressed. A star with an initial mass of about 8 times that of the Sun will turn into the heaviest possible white dwarf with the mass of 1.4 Suns. Degenerate objects with greater masses, beyond this threshold known as the Chandrasekhar limit, are too heavy to be supported by degeneracy pressure and cannot exist in stable equilibrium. At the other end of the stellar mass range, stars with initial masses of about one tenth of the Sun become white dwarfs with almost the same mass.

Brown dwarfs represent the second most numerous type of degenerate remnant. These bodies are stellar-like objects that are born with too little mass to sustain hydrogen fusion. Stellar objects with masses smaller than about 8% of the Sun fail to produce sustainable nuclear fusion and must live out their lives as brown dwarfs. The star formation process, as it makes stellar bodies, does not distinguish between objects above and below the hydrogen burning threshold. As a result, galaxies make substantial numbers of these failed stars. Although astronomers are still making a census of the stellar population at the low end of the mass spectrum, current estimates suggest that about 1 out of every 4 stellar objects is born as a brown dwarf. These objects do very little and hence are still around in the future. In particular, since they cannot sustain nuclear reactions, the brown dwarf population represents an important source of hydrogen fuel for the future universe.

Neutron stars are yet another type of degenerate objects. The largest progenitor stars, about 3 or 4 out of every 1,000, do not turn into red giants and then white dwarfs after they run out of nuclear fuel. They have a much more dramatic end. These massive stars, with more than 8 times the mass of our Sun, first burn their hydrogen into helium, and then burn the helium into carbon and oxygen. The process keeps going until they burn the material in their central cores all the way to iron. Of all the elements, iron represents a special benchmark in nuclear physics: Iron nuclei have the highest binding energy. As a result, one gets energy out of processes that fuse lighter elements into larger ones, as long as the products are smaller than iron. On the other hand, one can also get energy from breaking apart large nuclei – such as uranium – as long as the products are larger than iron. When a massive star burns its central core to become primarily iron, it can no longer squeeze any more energy out of nuclear reactions. Without an energy source, the central pressure drops precipitously and the entire star collapses. The usual results of these events are violent supernova explosions, which release as much energy as an entire galaxy, but only for a few seconds. In the center of the explosion, nuclei are pressed together tightly, and a degenerate neutron star is produced. In these objects, the density is so large, about one quadrillion times the density of water, that the nuclei are in a close-packed configuration. The star, with roughly the mass of the Sun and a radius of only 10 km, becomes essentially one gigantic nucleus, with most of the electrons and protons combining to make neutrons. Hence the name "neutron stars". These exotic objects play a relatively minor role in the far future due to their diminutive numbers.

Finally, the most massive stars sometimes leave behind stellar remnants with masses greater than the Chandrasekhar limit. In such cases, the central remnant can collapse to form a black hole. This process is just now being studied by astrophysicists. The first clear evidence for stellar black holes has come in the last decade, and observations of highly energetic supernova-like explosions have started to provide additional data. Although this issue remains under study, only those stars with masses more than about 40 times that of the Sun are expected to leave behind black holes. In spite of their small numbers, these bodies are important, as the black holes will outlast the brown dwarfs, white dwarfs, and neutron stars by a large temporal margin (see the discussion in Sect. 3.6).

3.4.5 White Dwarfs Capture Dark Matter

White dwarfs are like dying embers from a fire – they have no intrinsic source of power. Instead these stars retain heat from their fiery past, and this energy slowly leaks out as they cool and fade away. In the absence of any external influence, white dwarfs would fade away on time scale comparable to the current age of the universe, about 10 billion years. In the far future of the Degenerate Era, when the universe is many thousands of times older, isolated white dwarfs would not have much impact on the energy budget. But white dwarfs are not isolated.

All of the stars, or the stellar remnants, in the galaxy are embedded within a vast halo of Dark Matter particles. These mysterious entities are thought to interact only through gravity and the weak nuclear force. They make up the majority of the mass of galaxies, so their gravitational influence determines orbits and other dynamical properties in the galaxies of today. In the far future, after the stars have condensed into degenerate remnants, the weak interactions of these dark particles play an important role.

As Dark Matter particles trace their orbits through the halo, they encounter white dwarfs and other stellar bodies. In fact, the particles are deflected toward the stellar remnants because of their strong gravitational fields, in a process known as gravitational focusing. In any case, a steady stream of Dark Matter particles flows through the white dwarfs and some fraction of them are captured, so they remain bound to the star. The particles subsequently fall into the central cores where their population gradually builds up. When a sufficiently high density of particles populates the core, they can interact through the weak force and annihilate with each other. The mass energy of the particles is thus converted into radiative energy. Soon, the system reaches a steady state, so that the rate of particle annihilation inside the star is in balance with the rate of particle captured from the halo.

Through this process of Dark Matter capture and annihilation, white dwarfs are endowed with an additional energy source. In approximate terms, the power output of a white dwarf is measured in units of quadrillions of Watts. This number is enormous, but we can put it in perspective as follows: The amount of power that Earth intercepts from the Sun, at the present time, corresponds to about one hundred quadrillion Watts. The white dwarfs of the future will have internal power sources that are comparable to the external power source of our Earth today. Adding to this analogy, the radial size of a white dwarf is comparable to the radius of the Earth. The surface temperature of the star will be a cool 70 K, cold enough to freeze air into a liquid and hence somewhat below the 290 K temperature of the Earth. Nevertheless, these stellar surfaces will be some of the hottest places in the cosmos of the Degenerate Era.

These white dwarfs of the future will play the role now filled by stars in the universe of today. They will be the standard sources of power, and every galaxy will have billions of them. Although these degenerate stars will be far dimmer than the stars of today, they can maintain their energy production far longer. Whereas solar-type stars "only" last for ten billion years, and the longest-lived red dwarf stars last for "only" tens of trillions of years, these white dwarfs can continue shining for nearly one hundred billion billion years. After this span of time, however, the white dwarfs are susceptible to being scattered beyond the galaxy as it evaporates. On somewhat longer time frames, the supply of Dark Matter will be exhausted, and this mode of energy generation will come to an end.

3.4.6 When Degenerates Collide

In the galaxy today, collisions between stars are rare. Some collisions take place within the cores of the densest stellar clusters, but the galaxy as a whole is relentlessly empty and offers little support for such fireworks. Over the vast expanses of time available in the future, however, events that are rare today can sometimes take place. After the universe is old enough, when its burned out stars have condensed into degenerate remnants, these rare collisions start to matter.

The population of degenerate bodies includes roughly comparable numbers of brown dwarfs and white dwarfs. In this context, the brown dwarfs are most interesting because they are hoarding the most important stores of unburned hydrogen fuel. Like their true stellar brethren, brown dwarfs are born with a composition that is primarily hydrogen. After trillions upon trillions of years, when all the stars are gone, these brown dwarfs will still have their fuel supply. Being the ultimate stellar slackers, brown dwarfs wait around for half of eternity and do absolutely nothing.

Although collisions will be rare, they sometimes occur in the darkness of the future. As it turns out, most brown dwarfs are close to the mass threshold required to sustain nuclear burning of hydrogen. When two such bodies collide, the merger product is usually massive enough to become a true star. The typical stellar entity produced during the collision is about ten times smaller in mass than the Sun, and the resulting star will live for trillions of years. Since we know the number of brown dwarfs in the galaxy, the typical distribution and speed of their orbits, and the effective target area they present for colliding, we can estimate the rate at which small stars are made through this violent process. Such star forming events take place every few trillion years, so that only one or two stars at a time will be shining in the galaxy of the future. And these stars will be dim, with 1000 times less power than our Sun. This dark projection stands in sharp contrast to the present-day galaxy, where billions of stars are shining brightly.

In addition to the brown dwarfs, the galaxy contains even more white dwarfs, and these objects will collide with each other at similar rates. Most white dwarfs have small masses, and a collision between two small white dwarfs results in just another degenerate object. When the largest white dwarfs collide, however, the merger product can be larger than the maximum mass that a white dwarf can maintain. This mass threshold is known as the Chandrasekhar mass, named after the Indian astrophysicist who first realized its importance in stellar structure. When the Chandrasekhar mass limit is breached, thedegenerate star is too heavy to support itself, and the object blows up in a supernova. Through the action of these white dwarf collisions, the dark galaxies of the future will be punctuated by rare but spectacular explosions.

3.4.7 Proton Decay in Degenerate Stars

When the universe grows older than about 10^{25} years, galactic disks will have evaporated, stellar collisions will have run their course, and much of the Dark Matter supply will be used up or dispersed. In this bleak future, the remaining stellar remnants must turn to another source for their power. Over the span of time that are enormously longer than any, we have discussed thus far, protons are (probably!) unstable to decay and can provide an important energy source to the stars.

We think that protons should decay. All physical systems tend to seek lower energy states (water flows downhill) and the proton has a lower energy state available – the positron carries the same charge and 2,000 times less mass. The problem is getting there. In this example, the proton, which is made of matter, must transform itself into a positron, which is the anti-matter partner of the electron. This transformation involves what is known as "violation of conservation of baryon number", which is a type of process that has never been observed. However, we have good reason to think that such exotic processes must operate at some level. They must take place in order for the universe to be filled with matter rather than anti-matter (if the universe had both matter and anti-matter in significant abundance, they would annihilate with each other and produce enormous bursts of gamma radiation, but such explosions are not seen). In other words, the matter content of our universe implies that baryon number conservation must be violated and that protons must eventually decay.

The physics of proton decay is currently in an awkward state. On one hand, proton decay is in fact a science. Experiments to measure the lifetime of the proton are in their fourth generation (depending on how you count), so the field is not without experimental data. The problem is that the experiments run thus far have not actually seen protons decay. Instead, they have placed a strong lower bound on the proton lifetime. At the present time, only one experiment remains, the Super-Kamiokande observational facility in Japan, and its current data indicate that protons must live longer than 10^{33} years (Figs. 3.7 and 3.8).

In addition to these experiments, theoretical considerations suggest that the protons are expected to decay within 10^{45} years. On this longer time scale, virtual black holes are (likely to be) effective enough to drive proton decay, although our understanding of quantum gravity effects remains in its infancy. In any case, we have an experimental lower bound on the proton lifetime and a theoretical upper bound. In between, the proton lifetime can vary by a factor of nearly one trillion (10^{12}).

In spite of the reckless uncertainty in the proton lifetime, we can provide a good description of the long term fate and evolution of stellar remnants facing proton decay. Most of the mass is locked up in white dwarfs, so we focus on their evolution.

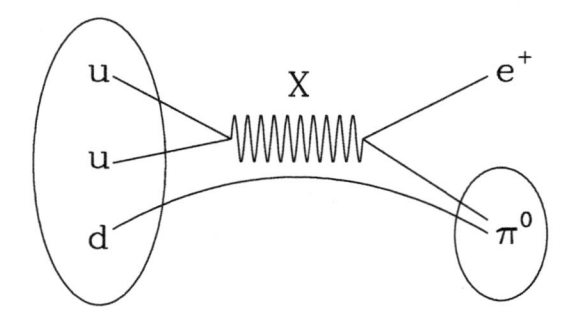

Fig. 3.7 Feynman diagram showing one typical channel for proton decay. The left side of the diagram shows the original proton, which is made up of three quarks. Two of the quarks interact through the intermediate particle (labeled with an X) and the proton is transformed into a positron (e^+) and a pion (π^0)

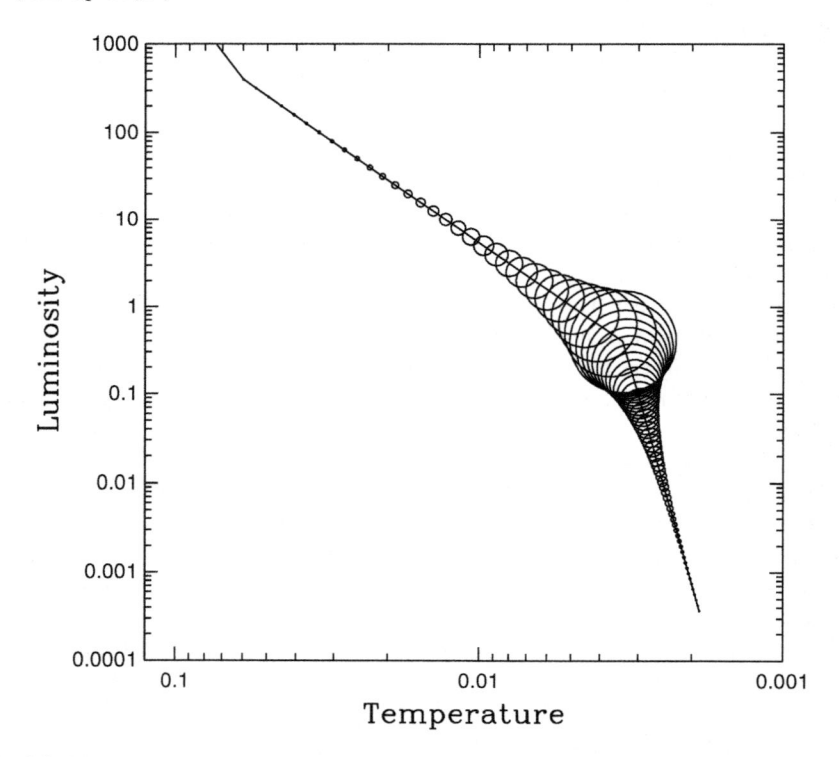

Fig. 3.8 Long term evolution of a white dwarf powered by proton decay. The vertical axis plots the power output of the star, given here in Watts. The horizontal axis plots the surface temperature of the star, given here in degrees Kelvin, and plotted backwards. The size of the symbols represent the radius of the stellar remnant as it loses mass. While the star remains degenerate, it grows larger as it loses mass; after it becomes about the size and mass of the planet Jupiter, the star shrinks as it loses more mass

In order to provide representative numbers, we assume here that the proton lifetime has an "intermediate" value of 10^{37} years. All of the particular numerical values can be rescaled for other choices.

While a white dwarf remains younger than the proton lifetime, its constituent protons still in fact decay. As in the case of radioactivity, protons do not sit around for these enormous spans of time and then all decay at once. Instead, at any given time, protons have a probability of decaying. Proton decay is thus taking place in white dwarfs today. At the present cosmological epoch, however, the amount of mass lost through this process, and the amount of energy generated, is completely inconsequential. The amount of heat leftover from the star's formative stages completely overwhelms that generated by proton decay.

In the far future, after the white dwarfs have exhausted all other means of generating energy, proton decay will become the dominant mechanism by default. A typical white dwarf with half the mass of the Sun will shine with the power of a few hundred Watts, so that entire stars are reduced to acting like light bulbs. These white dwarfs of the future will be far colder, however, with surface temperatures less than one degree on the Kelvin scale.

When the age of the white dwarf exceeds the proton lifetime, the effect of nucleon decay starts to change the structure of the star. As it loses mass, the stellar remnant actually grows larger in radius. This counterintuitive behavior occurs because the star is degenerate, where the pressure that supports it against collapse is provided by the quantum mechanical uncertainty principle. The white dwarfs starts with a stellar mass and the radial size of the Earth. It then loses mass and expands in size until it has the mass and the radius of our planet Jupiter. At this point in its future evolution, only one tenth of one percent of its original mass remains. The stellar remnant is no longer a degenerate white dwarf, but rather a huge block of hydrogen ice. As the "star" continues to lose mass, it now grows smaller in radius, until it fades into oblivion.

In the distant future this fate awaits for our Sun, as well as the vast majority of stars, which also become white dwarfs upon their death. The neutron stars and brown dwarfs will behave in similar fashion. Of all the stellar bodies in the universe, only the black holes are impervious to the destructive implications of proton decay. They slide unscathed through this epoch and live to generate energy over even longer spans of time.

3.5 The Future of Planets

Although planets contribute relatively little to the mass budget of the universe, and they provide essentially no power, these bodies are nonetheless important: Planets are the only places within the universe where life, as we know it, can arise. Of course, since we have no working theory for the origin of life, and little understanding of biogenesis in general, life could evolve in vastly different cosmic environments. However, biology in such alien settings would not be the kind with which we are familiar. We also note that "planets" could also be "moons", large rocky bodies that orbit about larger planets. In any case, the future of planets, broadly defined, constrains the future of life, at least as we know it.

3.5.1 Death by Red Giants

One class of planets that captures our interest are those that reside in "habitable orbits", locations within a solar system where the surface temperature allows for water to exist in liquid form. This location varies with the mass of the parental star, since the brightness of these celestial furnaces depends sensitively on the stellar mass. Stars larger than the Sun are much brighter than our modest star, but these larger cousins live for much shorter times. As a general rule, only stars with initial masses smaller than about two Suns can live long enough for life to develop, where we assume that at least one billion years is required. Stars of this mass, and smaller, do not explode at the end of their hydrogen burning lives. Instead, many of them swell up into red giants, and then slowly fade away as white dwarfs.

For planets living near aging stars, especially planets in habitable orbits, the red giant stage of stellar evolution can be dangerous. We have already considered how the red giant phases of the Sun will envelop Mercury, then Venus, and finally Earth. Stars with comparable masses end their lives in analogous fashion and will consume any terrestrial planets residing in the inner portions of their solar systems. For these solar type stars, the location of habitable zones roughly coincides with the extent of the inner solar systems that are destroyed during the red giant phases.

Small stars are different. For the stars in the smallest class, the red dwarf stars, prospects for habitability are altered. During the majority of their hydrogen burning lives, these stars are so dim that any habitable planets must be ten times closer than Earth is to the Sun. Like all stars, red dwarfs grow brighter as they get older, especially after a significant fraction of their hydrogen has been processed into helium. Unlike larger stars, however, these dwarfs do not become red giants during their death throes. Instead, they stay roughly the same size and grow somewhat bluer as they become more luminous. And like everything they do, these small stars become bluer at a leisurely pace. These late blue dwarf phases will not occur until the universe is about one thousand times older than today. At this late epoch, long after solar type stars have faded and shriveled to become white dwarfs, a star with 16% of the Sun's mass will experience a 5 billion year span with roughly constant power output. And its brightness will be nearly one third than that of the present-day Sun. Any favorably situated planets can – in principle – come out of cold storage and potentially become habitable.

3.5.2 Destruction by Supernovae

When a supernova is detonated in the core of a massive star, any planets living in its solar system are in grave danger. Near the exploding star, in the inner solar system where any habitable planets might conceivably reside, the blast from the supernova consists primarily of the ejected stellar material. As a rough picture of the blast, the supernova ejects about one solar mass worth of material at the fantastic speed of 10,000 km/s. With this high speed, about 1/30th of the speed of light, the ejected material carries approximately 10^{51} ergs of energy, the typical explosive energy rating for a supernova.

These ejection speeds are far greater than the escape speed from the surface of a star, or a planet, and are also far greater than the thermal motions of particles in any kind of familiar substance. As a result, any planets that have the misfortune to lie in this line of fire will experience the supernova blast as a rain of protons, about 10^{57} of them, hurtling through space and acting as microscopic bullets. These bullets, in turn, enforce destruction upon objects in their path.

Consider the case of an unfortunate planet that lives within the habitable zone of a massive star (although such a planet is unlikely to be inhabited because of the star's unfavorably short lifetime). When the star explodes as a supernova, the impact energy reaching an Earth-like planet would be 10 million times greater than that of the famous K/T comet that killed our dinosaurs. If you stood on this hypothetical planet, and faced into the blast, your body would be exposed to approximately one metric ton of stellar material, all moving at 1/30th the speed of light. Any ordinary life form would be immediately destroyed by this cosmic onslaught. But what happens to planet itself?

The distance of the habitable zone is far enough away from the star to keep the surface temperature of the planet within the range of liquid water. At this location, the planet would suffer severe damage from the supernova blast wave, but complete annihilation would be avoided. The planet would absorb about 10^{20} g of new material, the mass of a large asteroid, but 10 million times less than the mass of Earth. Even with the large impact velocity, the planet would gain a radial speed of only 13 cm/s. The planetary surface, however, would be blasted away and rearranged during the impact. The incoming protons have about the same kinetic energy as those found in nuclear reactions. In order for this enormous energy to be absorbed and redistributed, the rock would change its structure down to a mixing depth of hundreds of kilometers. The outer layers of the planet would become unrecognizable. In spite of the trauma, the planet itself would survive.

3.5.3 Long Term Fate of Planets

Planets that are too close to their stars can be destroyed by the red giant phase of solar type stars or by supernova blasts from larger stars. Many planets will survive, however, and will continue to orbit about stellar remnants long after the stars have ceased their nuclear operations. Still other planets can be produced *after* a supernova explosion, by condensing from the disk of leftover debris that orbits the resulting neutron star. What is the fate of these surviving astronomical bodies?

The possibility that Earth can be scattered out of our solar system was discussed previously. The odds of such an event are low, only about 1 part in 10^5 over the next 3 billion years. If we consider much longer spans of time, the odds increase dramatically. Generally speaking, the odds of Earth ejection would become significant over a time scale that is 10^5 times longer, about one third of a quadrillion years. All stars are thus susceptible to losing their planets on comparable time scales. By the time the universe is old enough for planet ejection to be important, however, the Sun and all the other stars will be degenerate stellar remnants. The time required for planet removal varies with the orbital radius, but most planets are expected to be liberated by the time the universe is about 10^{16} years old.

For planets that orbit too close to their stellar remnants, another fate lies in store. Orbits lose energy by emitting gravitational waves. These waves are produced by orbital motions, much like radio waves are produced by electron oscillations in radio transmitters. In the case of gravity waves, however, the process is highly inefficient. To illustrate this point, consider a planet in orbit about a solar type star, with an initial orbit of 1 AU (like the orbit of Earth). The planet will slowly spiral inward, and will be eventually accreted onto the central star, but only after nearly 10^{19} years have elapsed. For planets that have starting orbital radii as large as 1 AU, the scattering effects described above will act first. For extremely close planets, with orbital radii ten times smaller (0.1 AU), gravitational radiation will lead to their ultimate demise.

3.6 Black Holes

One non-technical definition of black holes is that they are "astronomical objects with gravitational forces so strong that not even light can escape their surfaces". In the far distant future, after proton decay enforces the demise of all other stellar bodies, black holes will be the brightest objects in the night sky. They will shine with radiation produced by the Hawking effect, a slow quantum mechanical process that ultimately leads to the decay of even the black holes.

The universe of the future is thus powered by Hawking radiation, named after physicist Stephen Hawking, who first discovered this process. Although the mathematical details are complicated, beyond the scope of this chapter, one can nonetheless get a conceptual picture of how black holes can radiate energy. Before doing so, however, let's take a look at the types of black holes that are expected to be present in the universe of the future.

3.6.1 Inventory of Black Holes

Galaxies produce black holes of two basic varieties. The first type of black hole, called a stellar black hole, results from the death of a massive star. Such massive stars are rare. Only three or four out of a thousand stars have enough mass to explode at the end of their lives in a supernova. And only a small fraction of supernova explosions are expected to produce black holes. Given the large number of stars formed in a galaxy over the course of its life, we expect a large galaxy like the Milky Way to produce approximately one million stellar black holes. These objects have masses in the range of ten to perhaps one hundred Suns.

The second type of black hole is produced in the centers of galaxies. Although the mechanism(s) that make such black holes are still being worked out, astronomical observations clearly show that almost every large galaxy contains a monster black hole at its core. These black holes come in a range of masses, from about one million to one billion times the mass of our Sun. Because of these enormous masses, after much careful consideration, astronomers have named this type of object "supermassive black holes".

The cosmic inventory of black holes is thus rather simple. Each large galaxy, like our Milky Way, contributes one supermassive black hole and roughly one million stellar black holes to the cosmic supply. And when the universe becomes old enough that proton decay has run its course, these black holes are the only stellar objects left in the universe.

These two "types" of black holes are different only in their masses. The supermassive black holes are much larger than their stellar counterparts. Except for their masses, however, all black holes are basically the same. More specifically, black holes are subject to a constraint known as the "No Hair Theorem", which states that only three properties of a black hole can be observed outside its event horizon. The first and most important observable property is the black hole mass. As outlined above, astrophysical black holes display a wide range in mass, from a few times the mass of the Sun up to billions of Suns. The second property is the spin of the black hole. Although both the current observational data and the available theory are preliminary, they indicate that black holes tend to spin rapidly, close to their maximum allowed values. The third and final property that could be observed is the electric charge of the black hole. In astrophysical settings, however, positive and negative charges tend to cancel out on the large mass scales that comprise both stellar and supermassive black holes. As a result, charge is not expected to play an important role in the black holes of the future. Mass and spin are the only characteristics that matter to black holes.

For completeness, we note that a third "type" of black hole is possible: Many authors have speculated that physical processes in the early universe could produce black holes with tiny masses. However, these primordial black holes, as they are often known, have never been observed. In addition, such black holes are expected to have small masses, perhaps comparable to elementary particles, and would evaporate too quickly to survive to the present epoch.

3.6.2 Hawking Radiation

Given that Hawking radiation runs the universe of the far future, it is useful to see how it works. In order to understand the Hawking effect, one has to keep three processes in mind at the same time. First, gravitational fields depend on the distance from a massive object, in this case the black hole. Closer to home, as you read this passage, your feet are (most likely) closer to the center of Earth than your head. As a result, the gravitational pull of the Earth is stronger on your feet, and your body is being stretched apart. This effect is called a tidal force. In analogous fashion, the moon pulls on the oceans and causes the tides. On the surface of Earth, this tidal force is relatively weak and hence is of little consequence on small scales. Near the event horizon of a black hole, however, the tidal forces can be truly enormous. After all, black holes are extreme objects and have extreme effects, in this case, extreme tidal forcing.

The next part of the Hawking process involves a counterintuitive result from quantum mechanics. Because of the Heisenberg uncertainty principle, the energy of empty space is not precisely empty. Instead, particles tend to burst in and out of existence. As a rule, particles cannot be created out of nothing, because doing so would require the creation of energy, and the laws of physics do not allow such behavior. We say that energy, including mass energy, is conserved. Nevertheless, such instances of particle production can and do occur, as long as the accompanying violations of conservation of energy take place only for a short time. The allowed time intervals are constrained by the quantum mechanical uncertainty principle. The larger the amount of energy found in the created particles, the shorter their allowed lifetimes. As a result, such virtual particles, as these entities are known, are continually created and soon after destroyed.

The final ingredient is a basic concept from elementary physics, namely that the work done on an object is equal to the force acting on it multiplied by the distance over which the force acts. Putting the three processes together, a simple description of the Hawking effect emerges: Through quantum effects, virtual particles are created near the event horizon of a black hole. Although such particles only live for a short time, the tidal stretching force, which is enormous near a black hole, does work on them while they remain in existence. If the work done on the particle – by the tidal force – is large enough, the particle is promoted from virtual existence to "real" existence. The particle can then leave the black hole, and is thus effectively emitted by the hole. In this manner, black holes radiate energy, albeit at an extremely slow rate.

Next we note that the tidal force is due to gravity, and the fundamental source of gravity is mass. As a result, the mass of the black hole must decrease a little bit to pay for the expenditure of energy required to make the particles. In other words, black holes must lose mass in order to emit radiation through the Hawking effect. Thus, through the combination of strong gravity and quantum mechanics, black holes are not forever.

Although the Hawking process is glacially slow, in the long term black holes are expected to evaporate all of their mass energy through this mechanism. How long does it take? A black hole with the mass of our Sun will evaporate in 10^{65} years. A typical stellar black hole, produced by the death throes of a massive star, is expected to have the mass of ten Suns and will evaporate in 10^{68} years. The supermassive black holes that anchor the centers of galaxies live much longer. A black hole having one million solar masses, much like the one in the middle of our Milky Way, will last for about 10^{83} years. The largest black holes observed – indirectly of course – have the mass of a billion Suns and will live for 10^{92} years. And finally, to emphasize that black holes are not forever: If you could take all of the mass contained within the current cosmological horizon, and somehow condense that gigantic collection of mass and energy into a single black hole, the resulting object would live for "only" 10^{131} years. As a result, all black holes are expected to eventually evaporate, and leave behind only their radiative decay products.

The radiation emitted by a black hole is close to thermal equilibrium, and can be characterized by a temperature. However, the effective temperature of a large black hole is extremely cold, only about one-ten-millionth of a degree for objects with the mass of the Sun. As black holes lose mass, however, they heat up so that both their temperature and power output increase with time. This buildup starts slowly, but eventually accelerates: During the final second of a black hole's life, it radiates more than one million kilograms of mass, or, equivalently, about thirty million megatons of explosive energy. After waiting around for enormous spans of time, the black holes will leave the universe in dramatic fashion!

3.7 Summary

3.7.1 Five Ages of the Universe

This chapter has outlined the long term fate and eventual destruction of the astrophysical objects that populate the universe – including clusters, galaxies, stars, planets, and black holes. Each type of object goes through a life cycle, with a formation event, a time span of active evolution, and finally a death-like conclusion. The life of the cosmos itself follows an analogous trajectory, and this story can be organized into five main epochs.

In the beginning, just after the big bang event, the universe was too hot and energetic to contain any stellar bodies. All the matter content of the universe was in the form of individual particles. This time before the dawn of stars is sometimes called the Primordial Era, and is described in the previous chapters. This era begins at an age of about 10^{-43} s – the smallest slice of time that can be defined because of quantum gravity – and ends when the first structures condense out of the expanding universe. No structure can be formed until the radiation has decoupled from the matter, and this transition takes place when the cosmic age is about 300,000 years. The first stars are formed somewhat later, perhaps when the universe is one million years old, or shortly thereafter.

After the first stars form, the universe changes its character. At present cosmological epoch, stars provide most of the energy that is generated, and they will continue to shine in this leading role for trillions of years. We are thus in the midst of an age of stars, called the Stelliferous Era, when the energy is produced through thermonuclear fusion in stellar cores. This era began sometime after the universe was a million years old, and will continue for several trillion years. After this time, the galaxies will run out of gas to make new stars, and the existing stars will exhaust their supply of hydrogen fuel. With no more nuclear burning stars, the universe will change its character again, and grow much darker.

After the stars have burned out, the remnants they leave behind will be the primary stellar bodies in the universe. These entities include brown dwarfs, white dwarfs, neutron stars, and black holes. Since the first three types of remnants are supported by quantum mechanical degeneracy pressure, this future epoch is often called the Degenerate Era. During this future time period, the universe remains active: white dwarfs capture Dark Matter particles, brown dwarfs collide to make new stars, white dwarfs collide to ignite supernovae, galactic disks evaporate, and the Dark Matter halos eventually annihilate. This era draws to a close when the universe becomes old enough for protons to decay, at a future epoch more than 10^{33} years from now. All of the degenerate stellar remnants lose their mass through proton decay and eventually fade into nothingness.

Black holes are the only stellar objects that survive the dilapidation enforced by proton decay, and they inherit the universe during the subsequent Black Hole Era. In the darkness of this future epoch, the radiation produced by black holes through the Hawking mechanism powers the universe. As the black holes shine, they also lose mass, and must eventually evaporate. The largest black holes live the longest. But after 10^{100} years, all of them will have made their explosive exits, and the universe changes its character once again.

After the black holes are gone, no stellar objects of any kind are left to light up the skies. Only the leftover waste products from the previous eras remain, and the universe slides into its Dark Era. The cosmic inventory is now extremely sparse, containing electrons, positrons, neutrinos, Dark Matter particles, and photons of stupendously long wavelengths. In the Primordial Era, the universe contained no stars – only particles – because it was too hot and too young. In the DarkEra in our distant future, the universe again has no stars – and only particles – because it is too cold and too old. These endpoints frame the story of our universe: Instead of evolving from ashes to ashes, or dust to dust, the cosmic timeline runs from particles to particles.

3.7.2 Forever Is a Long Time

The story of the future universe presented in this chapter represents our best available projection, based on our current understanding of the laws of physics and the nature of the cosmos. As such, this future timeline is based on honest calculations, and the underlying equations that govern these processes are relatively well understood. The appendix to this chapter outlines some of the relevant physics that drives our cosmic future. Nevertheless, this picture of the future universe requires a great deal of extrapolation, and an appropriate dose of caution is in order.

Although we begin with laws of physics that are well-vetted through experiments, we must extrapolate these results – by necessity – to time scales far longer than the current age of the universe, and hence far longer than those of the experiments. One nagging issue is that the laws of physics could change over these vast expanses of time. As one example, the strength of gravity is set by a universal gravitational constant that we denote as G. Current experiments show that G does not change on time scale shorter than a trillion years. While this time is 100 times longer than the current age of the universe, it still remains short compared to the times relevant to the Five Ages outlined above. If the gravitational constant were to vary over longer time scales, beyond the experimental bounds, significant corrections to the future history could be required. In similar fashion, changes in other aspects of physical law could lead to corresponding changes in our future projections.

As a general rule, our powers of prediction grow weaker as we consider time scales that probe ever deeper into the future. Not only can the laws of physics change on such enormous time scales, but also additional physical processes, that we know about in today's universe, may play an important but unforeseen role. This situation is roughly analogous to the difficulties we face in studying the early universe, when the cosmic age must be measured in tiny fractions of a second. In this early realm, the relevant energy scales are far above those we can study in experiments, whereas the time and length scales are much smaller. However, the early universe has one important advantage over the future universe, in that primordial processes can leave behind signatures that we can observe today. Nonetheless, our confidence diminishes as we go back farther in time.

Returning to matters of the future universe, we note that one could contemplate physical processes that occur on time scales far longer than those considered here. As one example, cold fusion can convert all existing atomic nuclei into iron, but the time required is estimated to be $10^{1,500}$ years. Such enormous time scales are not considered in this treatment for two reasons. First, we felt that the longest time into the future that allows for predictions of any confidence is about 10^{100} years, roughly the time required for Hawking evaporation of the largest black holes. Second, as one considers effects that take place over even longer times, the chances increase that something else will take place first and supersede. In the case of cold fusion, for example, it is highly likely that the nuclei in question will undergo nucleon decay long before cold fusion is important.

3.7.3 Our Place in Time

This projection of our future universe may seem dark, not only literally but also figuratively. Indeed, the story does not have a typical Hollywood ending, as all of the characters die in the end. While some may find this fate disconcerting, considerations of our cosmic future can also be uplifting.

This description of our future history is based on the laws of physics. One cause for optimism is that we understand physical law well enough to make sensible projections. Construction of this future timeline thus celebrates the power of our knowledge. In addition, this description of our future history is possible because, in spite of its complexity, the universe has been kind to us: The laws of physics, and manner in which they are realized within our local patch of space–time, are much simpler than they could have been. For example, our local universe is homogeneous and isotropic, which allows the cosmic expansion to be described using the simplest version of the equations resulting from Einstein's theory of general relativity. One can easily imagine alternative universes with more complex properties, and it would be much harder to build cosmological models under such circumstances.

The time scales involved in this story are fantastically large. Our universe will continue to live and evolve over vast tracks of time, perhaps far longer than the 10^{100} year cut-off considered here. Along the way, the cosmos will explore a wide variety of mechanisms for generating energy and entropy, including thermonuclear fusion in stellar cores, Dark Matter capture in degenerate white dwarfs, proton decay in stellar remnants, direct particle-on-particle annihilation, and Hawking radiation from black holes. At the present cosmological epoch, with the current age of only 14 billion years, our universe is still very much in its infancy. Even our current Stelliferous Era, when the universe is brightly lit, is near its beginning. Most of the stellar evolution that will occur in the universe belongs to our cosmic future, rather than the past. Stars will continue to provide power to the universe for trillions or even tens of trillions of years. As a result, most of the nuclear, chemical, geological, and even biological activities in our universe are yet to come.

Appendix: Quantitative Considerations

Regarding the Future History of the Universe

In this appendix, we outline some of the basic equations that describe the astrophysical processes related to the future history of the universe. The main text tells the story of our cosmic future; this additional discussion elucidates some of the relevant physics. Keep in mind, however, that this treatment itself remains approximate, roughly speaking, at the level of an intermediate undergraduate textbook.

This discussion is organized by the scales of the cosmos considered in the main text, from the universe on the largest scale down to black holes. We use cgs units (centimeters-grams-seconds) unless otherwise noted. The fundamental constants of physics are denoted by their usual symbols: gravitational constant G, reduced Planck constant $\hbar$, Boltzmann constant k, Stefan–Boltzmann constant σ, and speed of light c.

A.1 The Universe as a Whole

The expansion of the universe is given by Einstein's theory of relativity. This expansion can be measured by a scale factor $a(t)$ that tracks how much, and how fast, the universe expands. For the case of a universe that is both homogeneous and isotropic, the scale factor $a(t)$ of the universe obeys the equation of motion

$$\left(\frac{\dot{a}}{a}\right)^2 = \frac{8\pi G}{3}\rho_C \left(\Omega_{M0} a^{-3} + \Omega_{R0} a^{-4} + \Omega_V\right), \tag{3.1}$$

where ρ_C is the critical density, and where Ω_{M0}, Ω_{R0}, and Ω_V specify the fraction of the energy density in the form of matter, radiation, and the vacuum, at the present cosmological epoch, respectively. This equation assumes that the universe is spatially flat, thereby eliminating a fourth possible term. The universe is observed to be flat, to about 1%, so this approximation is valid. This assumption also requires the sum of the contributions to equal unity. The radiation contribution is negligible at the present time ($\Omega_{R0} \sim 10^{-4}$), so that most of the energy density of the universe is in the form of matter ($\Omega_{M0} \approx 0.3$) and the vacuum ($\Omega_V \approx 0.7$). Thus, $\Omega_{M0} + \Omega_V = 1$, as indicated by current experiments. In principle, the vacuum energy density, codified in the parameter Ω_V, could vary with time; however, observations suggest that Ω_V is nearly constant, and we consider only this case here.

In the future, as the scale factor $a(t)$ grows, the relative contribution of the matter terms grows smaller, and the vacuum term completely dominates the dynamics. This trend can be illustrated by the solution to (3.1) with only matter and vacuum components:

$$a(t) = \left(\frac{\Omega_{M0}}{1-\Omega_{M0}}\right)^{1/3} \left\{\sinh\left[\frac{3}{2}\sqrt{1-\Omega_{M0}}\, H_0 t\right]\right\}^{2/3}, \tag{3.2}$$

where $H_0 \approx 70$ km s^{-1} Mpc^{-1} is the Hubble constant. At later times, the scale factor approaches the simpler asymptotic form

$$a_{\text{asym}}(t) = \left(\frac{\Omega_{M0}}{4\Omega_V}\right)^{1/3} \exp\left[\sqrt{\Omega_V}\, H_0 t\right], \tag{3.3}$$

which shows that the scale factor of the universe will soon expand exponentially (for constant Ω_V, as assumed here). One can also see from these equations that time dependence of the vacuum energy density can lead to variations in the predictions of the future expansion history.

Provided that the universe continues to accelerate with nearly constant vacuum energy density, the space–time will soon exhibit a constant horizon size scale given by

$$r_H = \frac{c}{H_0} \frac{2}{\pi} \left(\frac{15}{\Omega_V} \right)^{1/2} \approx 12,600 \text{ Mpc}. \tag{3.4}$$

This horizon scale sets the effective size of the universe for "microphysics", with the space–time of the entire "universe" extending to much larger scales. This scale also sets the wavelength of the background heat bath of the future universe.

One of the more speculative – but potentially important – possibilities is that the future universe could experience a phase transition. In this event, the vacuum state of the universe would tunnel from its current state of (high) energy density to one with lower energy density. In the simplest case, the probability P per unit volume per unit time for such a transition to occur is given by the formula

$$P = K \exp[-S_4], \tag{3.5}$$

where S_4 is the "four-dimensional Euclidean action" of the theory that describes the physics of the vacuum. This action, in turn, depends on the parameters of the theory. The important point is that the transition probability depends on the exponential of the action, so that the transition rate is extremely sensitive to the details of the theory. At the present time, we do not have a working theory – much less a definitive one – for the vacuum state of our universe. As a result, this transition probability is incredibly uncertain.

A.2 Galaxies

Galaxies continue to make new stars as long as they have a supply of unburned hydrogen gas, which provides the raw material. As the supply of gas grows lower, so does the rate of star formation. Observations show that the rate of star formation is directly proportional to the surface density of molecular gas in a galaxy. The molecular gas content depends on both the conversion of atomic gas into molecular form and any return of material from the stellar population back into the interstellar medium. In approximate terms, this process can be represented by a simple equation of the form

$$\Gamma_* = -\frac{dM_{\text{gas}}}{dt} = M_{\text{gas}}/t_{\text{sf}}, \tag{3.6}$$

where M_{gas} is the total (integrated) mass of gas. The time scale t_{sf} is comparable to the current age of the universe (about 10 Gyr). The time t_* required for the gas supply to be completely depleted and is thus given by

$$t_* = t_{\text{sf}} \ln \left[M_{\text{gas}}/1 M_\odot \right], \tag{3.7}$$

where we allow the final gas supply to fall to the mass of the Sun. The time t_* thus represents the age of the universe when the last star is made through conventional processes. For typical disk galaxies at the present epoch, the gas supply is about $10^{10} M_\odot$, so that the gas supply is expected to be depleted in about 20 – 30 Hubble times, or $t \sim 300$ Gyr. With extra gas falling on galactic disks from halos, and especially star formation rates that drop below the linear law assumed in (3.6), the time span for star formation can be extended to perhaps one trillion years; after that time, star formation through ordinary channels will cease.

The galaxy adjusts its structure through the process of dynamical relaxation. The time scale for relaxation is given by

$$t_{\text{relax}} \approx \frac{R}{V} \frac{N}{10 \ln N}, \qquad (3.8)$$

where R is the system size, the speed V represents the random motions of the stars, and N is the number of stars. This scale t_{relax} represents the time required for a typical star to change is velocity significantly (by a factor of order unity). The time required for the disk of the galaxy to evaporate most of its stars is about 100 times longer than the relaxation time. As a result, the time over which the stellar disk of the galaxy remains intact is expected to be about $10^{19} - 10^{20}$ year.

Dark Matter halos can evolve by annihilation of their constituent particles. Such annihilation events can occur in direct particle-on-particle interactions and also indirectly through capture by stellar remnants (see the following subsection). For the direct process, the interaction rate takes the form

$$\Gamma_{\text{dm}} = n_{\text{dm}} \langle \sigma_{\text{dm}} v \rangle, \qquad (3.9)$$

where n_{dm} is the number density of Dark Matter particles, σ_{dm} is the cross section for annihilation, and v is the typical particle velocity. The interactions take place through the weak force, and the cross section takes the approximate form

$$\sigma_{\text{dm}} \sim 5 \times 10^{-34} \text{ cm}^2 \left(\frac{m_{\text{dm}}}{100 m_P} \right)^2, \qquad (3.10)$$

where m_{dm} is the mass of the Dark Matter particles and m_P is the proton mass. The cross section is scaled to 100 proton masses because the current data suggest masses of this order. If we take the number density to be $n_{\text{dm}} = 0.01$ cm^{-3}, the characteristic time scale, given by $t_{\text{dm}} = 1/\Gamma_{\text{dm}}$, becomes $t_{\text{dm}} \approx 3 \times 10^{21}$ years. The galactic halo does not exhibit a single density, with the outer parts of the halos having more mass and lower density. The time required for the entire halo to be compromised is thus somewhat longer than this estimate. In addition, given that Dark Matter particles have not yet been detected and hence their properties remain unverified, these numbers should be regarded with appropriate caution.

A.3 Stars

For a star of a given mass, detailed stellar evolution models can be used to self-consistently determine its radius, density structure, temperature structure, and luminosity, all as a function of time. Here we adopt a simpler approach. The nuclear burning lifetime $t_{\rm nuc}$ is an important stellar property, and can be written in the form

$$t_{\rm nuc} = E_{\rm nuc}/L_* = \frac{f_c \epsilon M_* c^2}{L_*}, \qquad (3.11)$$

where L_* is the luminosity and $E_{\rm nuc}$ is the total energy available in nuclear fuel. In the second equality, we evaluate this energy supply in terms of the stellar mass M_*, the fraction f_c of the star that cycles through the core, and the nuclear efficiency factor $\epsilon = 0.007$ for converting hydrogen into helium. For solar-type stars, only 10% of nuclear material is accessible ($f_c = 0.1$), and the stellar lifetime $t_{\rm nuc}$ is about 10 billion (10^{10}) years. A star with the mass of 16 Suns has a luminosity that is about 10,000 times brighter than the Sun. Such a star will live for only about 10 million years. On the other end of the mass range, a red dwarf with only one tenth of a solar mass will live for about 6 trillion years.

Over sufficiently long time scales, new stars can be made through collisions of brown dwarfs. The collision rate is given by the usual formula

$$\Gamma_{\rm bd} = n_{\rm bd} \sigma_{\rm bd} v, \qquad (3.12)$$

where $n_{\rm bd}$ is the number density of brown dwarfs, v is their relative velocity, and $\sigma_{\rm bd}$ is the cross section for merging. The density of brown dwarfs is roughly similar to that of stars, i.e., about 1 per cubic pc in the galaxy today. Numerical simulations show that the effective target area of a brown dwarf for collisions is roughly comparable to its geometrical cross section, so one can use

$$\sigma_{\rm bd} \approx \pi R_{\rm bd}^2 \approx 3 \times 10^{20} {\rm cm}^2. \qquad (3.13)$$

The corresponding time scale $t_{\rm bd}$ for a typical brown dwarf to experience a collision is given by $t_{\rm bd} \sim 1/\Gamma_{\rm bd} \sim 10^{21}$ years. In the absence of other effects, the brown dwarf populations would be depleted via collisions and mergers on this time scale. However, this time is longer than the evaporation time of the galaxy, so that the collision rate will become lower as the galaxy spreads out. The collision rate for the entire galaxy is roughly $N_{\rm bd}/t_{\rm bd}$, which corresponds to one brown dwarf collision every 100 billion years. Since the merger product of these collisions is generally a small star, one that will live for perhaps a trillion (10^{12}) years, the galaxy is expected to have several such stars shining at a given time in the future.

In the future, stellar remnants play the role now filled by stars, and white dwarfs will be the dominant type. These objects are supported by electron degeneracy pressure, which enforce a mass–radius relationship of the form

3 The Future History of the Universe

$$R_* M_*^{1/3} \sim \frac{\hbar^2}{2cG} \left(\frac{Z}{Am_P}\right)^{5/3}, \tag{3.14}$$

where m_e is the electron mass and m_P is the proton mass. The white dwarf is assumed to have uniform composition, a single element with atomic number Z and atomic weight A. Equation (3.14) is remarkable in several respects. The right hand side of the equation is made of fundamental constants of nature, most of which describe physics on microscopic scales. In contrast, the left hand side of the equation contains stellar properties (mass and radius) that occur on enormous scales compared to those of everyday life. Another curious feature of this relation is that white dwarfs with larger masses have smaller radii. This behavior is in sharp contrast to ordinary experience: rocks with larger masses have larger sizes (radii), not smaller sizes.

In the future, white dwarfs and other remnants are the most important stellar constituents of the cosmos. After ordinary hydrogen burning stars are gone, white dwarfs provide an important power source by burning Dark Matter particles. For this process to take place, the stellar remnant must capture the particles. The optical depth of a white dwarf to a Dark Matter particle is given by $\tau = N_{\text{col}} \sigma_{\text{dm}}$ where N_{col} is the column density of the star and σ_{dm} is the cross section of interaction between Dark Matter particles and ordinary nuclei. The column density is determined by the structure of the star, but can be approximated using the white dwarf mass–radius relation of (3.14). We thus obtain

$$N_{\text{col}} \sim \frac{M_*}{m_P R_*^2} \sim \frac{M_*^{5/3}}{m_P} \left(\frac{2m_e G}{\hbar^2}\right)^2 \left(\frac{Am_P}{Z}\right)^{10/3}. \tag{3.15}$$

With this column density and the expected (but, unfortunately, still unknown) interaction cross sections for Dark Matter, almost every particle passing through the bulk of the star will be captured. The rate of capture is then given by

$$\Gamma_{\text{cap}} = n_{\text{dm}} \sigma_{\text{wd}} v, \tag{3.16}$$

where σ_{wd} is the effective target area that the white dwarf offers to the Dark Matter particles. The capture rate is rather large, about 10^{23} particles per second, where we have used the values as in the previous section. The star is expected to reach a steady state, where the rate of annihilation in the central core is balanced by the capture rate of (3.16). The power output of the stellar remnant is thus given by

$$L_{\text{dm}} = \Gamma_{\text{cap}} m_{\text{dm}} c^2, \tag{3.17}$$

where we assume that the entire mass m_{dm} of the Dark Matter particle is converted into radiation. The resulting luminosity is approximately $L_{\text{dm}} \approx 3 \times 10^{22}$ erg/s, which is more than one quadrillion Watts. The escape speed from the surface of the white dwarf is high, about 3,000 km/s, which is about ten times the speed of the Dark Matter in the galactic halo. As a result, the white dwarf will offer a somewhat larger cross section for Dark Matter capture through gravitational focusing. This effect, in turn, can enhance the luminosity by a factor of 10 – 100. However, this luminosity depends on the location of the white dwarf in the halo (which affects the number density $n_{\rm dm}$), as well as the unknown properties of Dark Matter.

When the universe becomes older than $\sim 10^{30}$ years, proton decay becomes one of the most important physical processes in the universe. Unfortunately, the lifetime of the proton has not yet been measured. All we know is that current experimental results indicate that protons live longer than $\sim 10^{33}$ years. Adding to the uncertainty, proton decay can take place through a vast number of different modes or channels, with varying intermediate states.

In order to illustrate this process, we consider a highly simplified description. Proton decay takes place on the scale of quarks, the smaller particles that make up protons. Roughly speaking, two quarks within the proton interact with each other through a large virtual particle, and transform themselves into other particles, for example a positron and an antiquark. The antiquark pairs with the third quark, but that short-lived composite particle (called a pion) soon decays into radiation. The positron is left behind to carry the original charge of the proton. When this process takes place within a white dwarf, the positron quickly finds an electron to annihilate, and its mass energy is also converted to radiation. On the "large" scale, this entire reaction takes the form

$$p^+ + e^- \to e^+ + \pi^0 + e^- \to \gamma_1 + \gamma_2 + \gamma_3 + \gamma_4, \qquad (3.18)$$

where p is the proton, e is an electron, π is a pion, and the γ_j are the four gamma ray photons produced. The superscripts keep track of the charge on the particles (charge must be conserved in all such interactions). The net result of such a proton decay event is to convert the entire mass energy of the proton, and an electron, into high energy radiation. These photons will randomly walk their way out of the star, and will degrade to longer wavelengths (lower energy) as they do so. The rate of proton decay through this type of process is given by the seemingly simple equation

$$\Gamma_{\rm P} = \alpha^2 \frac{m_{\rm P}^5}{m_X^4}, \qquad (3.19)$$

where $m_{\rm P}$ is the proton mass, m_X is the mass of the intermediate virtual particle that drives this process, and α is a coupling constant that sets the overall rate. One problem is that we do not know the coupling constant α, although our theoretical understanding suggests it could be confined to a reasonably small range. A bigger problem is that we do not know the mass scale m_X, and the decay rate depends on this value to the fourth power. Given our current experiments, we think that m_X is larger than about 10^{16} protons. In addition, an upper limit to the possible mass of such particles is set by the Planck scale, where quantum gravity effects become important, and this scale is equivalent to the mass of about 10^{19} protons. We are thus left with a range of 1,000 for the mass m_X, which corresponds to a range of 10^{12} for the proton lifetime. In other words, protons are likely to decay within the range

$$10^{33} \text{ years} \leq t_P \leq 10^{45} \text{ years}. \tag{3.20}$$

The lower limit is firm; the upper limit is probable, but not rigorous.

In spite of the uncertainty in the proton lifetime, the behavior of white dwarfs experiencing proton decay is relatively simple. Suppose that protons decay at a rate Γ_P, as described through (3.19), where we leave the values unspecified. The mass of the white dwarf will decrease as its constituent protons decay according to the relation

$$M_{\mathrm{wd}}(t) = M_{\mathrm{wd}0} \exp\left[-\Gamma_P t\right], \tag{3.21}$$

where $M_{\mathrm{wd}0}$ is the starting mass of the white dwarf. Note that since the progenitor stars live for at most trillions of years, which is incredibly small compared to the proton lifetime, it does not matter whether we start the clock from zero, or from the epoch when the white dwarf first condenses. The luminosity or power output of the white dwarf can then be written in the form

$$L_{\mathrm{wd}} = f_\nu \Gamma_P c^2 M_{\mathrm{wd}}(t) = f_\nu \Gamma_P c^2 M_{\mathrm{wd}0} \exp\left[-\Gamma_P t\right], \tag{3.22}$$

where the factor $f_\nu < 1$ accounts for neutrino losses. Some proton decay events lead to neutrino production, but these ghostly particles do not interact with the stellar material and hence are essentially lost from the system. Finally, with this luminosity, the white dwarf mass radius relation of (3.14), and the standard outer boundary condition for a star,

$$L_* = 4\pi R_*^2 \sigma T_*^4, \tag{3.23}$$

we can determine the surface temperature:

$$T_{\mathrm{wd}} = \left(\frac{f_\nu \Gamma_P c^2}{4\pi\sigma}\right)^{1/4} \left(\frac{2m_e G}{\hbar^2}\right)^{1/2} \left(\frac{Am_P}{Z}\right)^{5/6} M_*^{5/12}. \tag{3.24}$$

As the mass decreases, the stellar surface cools down.

A.4 Planets

Planets can be stripped from their parental stars by passing stellar bodies. Since the galaxy is fantastically empty, this process is highly inefficient. The stellar entities involved will thus be stellar remnants by the time this process becomes important. The cross sections for planets to be removed from their solar systems can be calculated by using a larger ensemble of Monte Carlo N-body simulations. Because passing remnants in the field of the galaxy move at high speeds, $v \sim 100$ km/s, they can pass right through solar systems without fully removing planets. The cross sections are thus low, with $\sigma_{ej} \sim 34$ AU^2 for planets in orbits akin to Neptune. The estimated time t_{ej} for such planets to be ionized from their stars is given by

$$t_{ej} \approx \frac{1}{n_* \sigma_{ej} v}, \qquad (3.25)$$

where the number density of stars $n_* \sim 1$ pc^{-3}. Using the values quoted above, we find an ejection time of $t_{eject} \approx 4 \times 10^{14}$ years. The cross section scales (approximately) linearly with the radius a (semi-major axis) of the planetary orbit, $\sigma_{ej} \propto a$, in contrast to the geometric relation $\sigma_{ej} \sim a^2$. The planets that survive the red giant phases of their stars are expected to have $a \geq 1$ AU, and these bodies will be detached from their stars in about 10^{16} years.

Planets that remain in orbit will be affected on longer time scales by gravitational radiation. The motion of the planet around its star generates extremely weak disturbances in the background space–time and these waves drain energy out of the system. For a planetary orbit with initial radius a, the time required to spiral into the central star is roughly given by

$$t_{gr} = \frac{2\pi a}{v} \left(\frac{c}{v}\right)^5, \qquad (3.26)$$

where v is the orbital speed and c is the speed of light. If the star has mass M_*, the orbital speed is given by the expression $v^2 = GM_*/a$. If the star has mass $M_* = 0.5 M_\odot$, typical for white dwarfs, then a planet orbiting at $a = 1$ AU will spiral inward to the star in $t_{gr} \approx 10^{19}$ years.

A.5 Black Holes

Black holes are small in size but large in mass, at least compared to other astrophysical objects. The size of a black hole is not completely straightforward to define, because these objects distort the geometry of the space around them. Nonetheless, one well-defined quantity is the Schwarzschild radius, which we somewhat generally refer to as the radius or size of the black hole; this radius is given by

$$R_{bh} = \frac{2GM_{bh}}{c^2}. \qquad (3.27)$$

This quantity is also called the gravitational radius. For a black hole with the mass of the Sun, this radius is only 3 km. For comparison, the radius of Earth is approximately 6,400 km, so a stellar black hole is smaller than a city. The Schwarzschild radius scales linearly with the mass, so the black hole in the center of our galaxy, with its mass of 3 million Suns, has a radius of 10 million kilometers (10^{12} cm). The largest black holes we find in the cosmos today have the mass of billions of Suns, or about 10^{15} cm (roughly the radial size of a solar system).

3 The Future History of the Universe

Black holes emit energy through the Hawking mechanism, a quantum mechanical process that operates in highly curved space–time, such as just outside the event horizon of a black hole. This process is currently understood in semi-classical terms, and has not yet been experimentally verified. To leading order, the spectrum of radiation emitted through the Hawking mechanism takes the form of a blackbody, and thus has a well-defined temperature. The Hawking temperature of a black hole is given by the expression

$$T_{bh} = \frac{\hbar c^3}{8\pi k G M_{bh}}, \qquad (3.28)$$

where M_{bh} is the mass of the object. Note that this temperature is extremely small: $T_{bh} \approx 6 \times 10^{-8}$ K for a black hole with one solar mass. The corresponding power output (luminosity) of the black hole is then given by

$$L_{bh} = 4\pi R_{bh}^2 \sigma T_{bh}^4 = \frac{\hbar c^6}{15,360\pi G^2 M_{bh}^2}. \qquad (3.29)$$

Larger black holes (in mass) thus have smaller power ratings, so that large black holes live for much longer than smaller ones. The lifetime of a black hole, due to mass loss from Hawking radiation, takes the form

$$\tau_{bh} = \frac{5,120\pi G^2}{\hbar c^4} M_{bh0}^3 \approx 10^{65} \text{ year} \left(\frac{M_{bh0}}{M_\odot}\right)^3, \qquad (3.30)$$

where M_{bh0} is the starting mass of the black hole. Stellar black holes, produced by supernovae from massive progenitor stars, are expected to have masses of 10 – 100 $M_\odot$ and hence lifetimes of 10^{68}–10^{71} year. Supermassive black holes that live in galactic centers have masses in the range 10^6–10^9 $M_\odot$ and hence lifetimes in the range 10^{83} – 10^{92} year. In the text, we have used the round number of 10^{100} years to mark the end of the Black Hole Era. This value corresponds to the evaporation time for black holes with masses of $\sim 5 \times 10^{11}$ $M_\odot$, which is about the largest such object that our universe could make.

End Notes

This section provides a brief review of the various issues covered in the text: The notes given below present a description of the relevant issues and the corresponding references. The citations refer to the reference list that follows. Note that (for brevity) only representative references are given.

I. Introduction. Many textbooks provide an overview of modern cosmology [30]. This manuscript deals with the future of the universe, which has been covered in several review articles [2, 11, 20, 28, 43], as well as popular-level books [2, 3, 18]. This present work updates and builds upon these previous efforts.

II. The Future of the Cosmos. The future of the large scale structure of the universe can be calculated under the assumption that the Dark Energy content is known [8, 9, 39]. The return to an effectively steady universe has been emphasized by Krauss and Scherrer ([33]; see also Loeb [38]; Adams et al. [1]). Phase transitions in the background space-time are possible with the right potentials for scalar fields [14–16, 27, 36]. The discussion of Heat Death dates back to Clausius [12, 13]; in the context of modern cosmology, heat death has additional implications [2]. The concept of the Big Rip was introduced by Caldwell et al. [10].

III. The Future of Galaxies. The isolation of galaxies follows from the future of large scale structure (again, see Nagamine and Loeb [39], Busha et al. [8], Adams et al. [1], Krauss and Scherrer [32]). The upcoming collision with Andromeda has been simulated by several groups (see Cox and Loeb [17] and references therein). Evaporation of the galactic disk is covered in standard texts [7].

IV. The Future of Stars. The future evolution of solar type stars is described in most stellar structure textbooks (e.g., Hansen and Kawaler [23]). The smallest red dwarf stars, which live far longer than the current age of the universe, have only recently been considered [35]. The inventory of degenerate objects depends on the relationship between stellar progenitor mass and the mass of the remnant [46]. The capture of Dark Matter particles was first considered in the Sun [42], and later for white dwarfs in the future universe [2]. Proton decay has a long history, with the latest experimental results given by the Super-Kamiokande experiment, which provide lower bounds on the proton lifetime. Theoretical upper bounds are inferred from gravitationally induced proton decay processes [26, 47]. The effects of proton decay on stars are given in Feinberg [21], Dicus et al. [19], and Adams et al. [4].

V. The Future of Planets. Scattering of planets has been studied in many contexts, including the future of our Solar System (e.g., Laughlin and Adams [34]). Habitable zones are considered by numerous authors and most astronomy textbooks.

VI. Black Holes. A general overview of blackholes and their place in the universe in given by Thorne [45]. Black holes evaporate via the Hawking process (Hawking [24,25]; see also Bekenstein [6]). From an astronomical standpoint, black holes have only recently passed the threshold from being candidate objects to being detected objects (starting with Kormendy et al. [31]).

VII. Summary. In addition to the aforementioned review articles [2, 20, 28, 43], a number of authors have also discussed "big picture" aspects of the future universe, including the Anthropic Cosmological Principle [5], Copernican Principles [22], the return to a static universe [32], multiple universe [44], and prospects for life [20, 32].

References and Further Reading

1. F.C. Adams, M.T. Busha, A.E. Evrard, R.H. Wechsler, The asymptotic structure of space-time, Int. J. Modern Phys. A **12**, 1743 (2003)
2. F.C. Adams, G. Laughlin, A dying universe: the long term fate and evolution of astrophysical objects, Rev. Mod. Phys. **69**, 337 (1997)
3. F.C. Adams, G. Laughlin, *Five Ages of the Universe* (The Free Press, New York, 1999)
4. F.C. Adams, G. Laughlin, M. Mbonye, M.J. Perry, The gravitational demise of cold degenerate stars, Phys. Rev. D **58**, 083003 (1998)
5. J.D. Barrow, F.J. Tipler, *The Anthropic Cosmological Principle* (Oxford University Press, Oxford, 1986)
6. J.D. Bekenstein, A universal upper bound to the entropy to energy ratio for bounded systems, Phys. Rev. D **23**, 287 (1981)
7. J. Binney, S. Tremaine, *Galactic Dynamics* (Princeton University Press, Princeton, 1987)
8. M.T. Busha, F.C. Adams, A.E. Evrard, R.H. Wechsler, Future evolution of cosmic structure in an accelerating universe, Astrophys. J. **596**, 713 (2003)
9. M.T. Busha, A.E. Evrard, F.C. Adams, R.H. Wechsler, The ultimate halo mass in a ΛCDM universe, Mon. Not. R. Astron. Soc. **363**, 11 (2005)
10. R.R. Caldwell, M. Kamionkowksy, N.N. Weinberg, Phantom energy: dark energy with $w < -1$ causes cosmic doomsday, Phys. Rev. Let. **91**, 1301 (2003)
11. M.M. Cirkovic, Resource letter: physical eschatology, Am. J. Phys. **71**, 122 (2003)
12. R. Clausius, Ann. Physik **125**, 353 (1865)
13. R. Clausius, Phil. Mag. **35**, 405 (1868)
14. S. Coleman, The fate of the false vacuum: 1. Semiclassical theory Phys. Rev. D **15**, 2929 (1977)
15. S. Coleman, *Aspects of Symmetry* (Cambridge University Press, Cambridge, 1985)
16. S. Coleman, F. De Luccia, Gravitational effects on and of vacuum decay, Phys. Rev. D **21**, 3305 (1980)
17. T.J. Cox, A. Loeb, The collision between the Milky Way and Andromeda, Mon. Not. R. Astron. Soc. **386**, 461 (2008)
18. P.C.W. Davies, *The Last Three Minutes* (BasicBooks, New York, 1994)
19. D.A. Dicus, J.R. Letaw, D.C. Teplitz, V.L. Teplitz, Effects of proton decay on the cosmological future, Astrophys. J. **252**, 1 (1982)
20. F.J. Dyson, Time without end: physics and biology in an open universe, Rev. Mod. Phys. **51**, 447 (1979)
21. G. Feinberg, The coldest neutron star, Phys. Rev. D **23**, 3075 (1981)
22. J.R. Gott, Implications of the Copernican Principle for our future prospects, Nature **363**, 315 (1993)
23. C.J. Hansen, S.D. Kawaler, *Stellar Interiors: Physical Principles, Structure, and Evolution* (Springer, New York, 1994)
24. S.W. Hawking, Black hole explosions? Nature **248**, 30 (1974)
25. S.W. Hawking, Particle creation by black holes, Comm. Math. Phys. **43**, 199 (1975)
26. S.W. Hawking, D.N. Page, C.N. Pope, The propagation of particles in spacetime foam, Phys. Lett. **86 B**, 175 (1979)
27. P. Hut, M.J. Rees, How stable is our vacuum? Nature **302**, 508 (1983)
28. J.N. Islam, Possible ultimate fate of the universe, Quart. J. R. Astron. Soc. **18**, 3 (1977)
29. J.N. Islam, *The Ultimate Fate of the Universe* (Cambridge University Press, Cambridge, 1983)
30. E.W. Kolb, M.S. Turner, *The Early Universe* (Addison-Wesley, Redwood City, CA, 1990)
31. J. Kormendy, et al., Spectroscopic evidence for a supermassive black hole in NCG 4486B, Astrophys. J. **482**, L139 (1997)
32. L.M. Krauss, G.D. Starkman, Life, the universe, and nothing: life and death in an ever-expanding universe, Astrophys. J. **531**, 22 (2001)

33. L.M. Krauss, R.J. Scherrer, The return of a static universe and the end of cosmology, Gen. Rel. Gravitation **39**, 1545 (2007)
34. G. Laughlin, F.C. Adams, The frozen earth: binary scattering events and the fate of the solar system, Icarus **145**, 614 (2000)
35. G. Laughlin, P. Bodenheimer, F.C. Adams, The end of the main sequence, Astrophys. J. **482**, 420 (1997)
36. A.D. Linde, Decay of the false vacuum at finite temperature, Nucl. Phys. **B216**, 421 (1983)
37. A.D. Linde, Eternally existing self-reproducing chaotic inflationary universe, Phys. Lett. **175B**, 395 (1986)
38. A. Loeb, Long-term future of extragalactic astronomy, Phys. Rev. D **65**, 7301 (2002)
39. K. Nagamine, A. Loeb, Future evolution of nearby large-scale structures in a universe dominated by a cosmological constant, New Astron. **8**, 439 (2003)
40. D.N. Page, Particle transmutations in quantum gravity, Phys. Lett. **95 B**, 244 (1980)
41. D.N. Page, M.R. McKee, Matter annihilation in the late universe, Phys. Rev. D **24**, 1458 (1981)
42. W.H. Press, D.N. Spergel, Capture by the Sun of a galactic population of weakly interacting massive particles, Astrophys. J. **296**, 679 (1985)
43. M.J. Rees, The collapse of the universe: an eschatological study, Observatory **89**, 193 (1969)
44. M.J. Rees, Our universe and others, Quart. J. R. Astron. Soc. **22**, 109 (1981)
45. K.S. Thorne, *Black Holes and Time Warps: Einstein's Outrageous Legacy* (Norton, New York, 1994)
46. M.A. Wood, Constraints on the age and evolution of the galaxy from the white dwarf luminosity function, Astrophys. J. **386**, 539 (1992)
47. Ya.B. Zel'dovich, A new type of radioactive decay: gravitational annihilation of baryons, Phys. Lett. **59 A**, 254 (1976)

Appendix A
The Return of a Static Universe and the End of Cosmology

Lawrence M. Krauss and Robert J. Scherrer

Abstract We demonstrate that as we extrapolate the current CDM universe forward in time, all evidence of the Hubble expansion will disappear, so that observers in our "island universe" will be fundamentally incapable of determining the true nature of the universe, including the existence of the highly dominant vacuum energy, the existence of the CMB, and the primordial origin of light elements. With these pillars of the modern Big Bang gone, this epoch will mark the end of cosmology and the return of a static universe. In this sense, the coordinate system appropriate for future observers will perhaps fittingly resemble the static coordinate system in which the de Sitter universe was first presented.

Shortly after Einstein's development of general relativity, the Dutch astronomer Willem de Sitter proposed a static model of the universe containing no matter, which he thought might be a reasonable approximation to our low-density universe. One can define a coordinate system in which the de Sitter metric takes a static form by defining de Sitter spacetime with a cosmological constant Λ as a four-dimensional hyperboloid $\mathcal{S}_\Lambda : \eta_{AB}\xi^A\xi^B = -R^2$, $R^2 = 3\Lambda^{-1}$ embedded in a $5d$ Minkowski spacetime with $ds^2 = \eta_{AB}d\xi^A d\xi^B$, and $(\eta_{AB}) = \mathrm{diag}(1, -1, -1, -1, -1)$, $A, B = 0, \ldots, 4$. The static form of the de Sitter metric is then

$$ds_s^2 = (1 - r_s^2/R^2)dt_s^2 - \frac{dr_s^2}{1 - r_s^2/R^2} - r_s^2 d\Omega^2,$$

L.M. Krauss (✉)
Department of Physics, Case Western Reserve University, Cleveland, OH 44106, USA

Department of Physics & Astronomy, Vanderbilt University, Nashville, TN 37235, USA
e-mail: krauss@cwru.edu

R.J. Scherrer
Department of Physics & Astronomy, Vanderbilt University, Nashville, TN 37235, USA
e-mail: robert.scherrer@vanderbilt.edu

which can be obtained by setting $\xi^0 = (R^2 - r_s^2)^{1/2} \sinh(t_s/R)$, $\xi^1 = r_s \sin\theta \cos\varphi$, $\xi^2 = r_s \sin\theta \sin\varphi$, $\xi^3 = r_s \cos\theta$, and $\xi^4 = (R^2 - r_s^2)^{1/2} \cosh(t_s/R)$. In this case, the metric only corresponds to the section of de Sitter space within a cosmological horizon at $R = r - s$.

In fact de Sitter's model was not globally static, but eternally expanding, as can be seen by a coordinate transformation which explicitly incorporates the time dependence of the scale factor $R(t) = \exp(Ht)$. While spatially flat, it actually incorporated Einstein's cosmological term, which is of course now understood to be equivalent to a vacuum energy density, leading to a redshift proportional to distance.

The de Sitter model languished for much of the last century, once the Hubble expansion had been discovered, and the cosmological term abandoned. However, all present observational evidence is consistent with a ΛCDM flat universe consisting of roughly 30% matter (both Dark Matter and baryonic matter) and 70% Dark Energy [1–3], with the latter having a density that appears constant with time. All cosmological models with a nonzero cosmological constant will approach a de Sitter universe in the far future, and many of the implications of this fact have been explored in the literature [4–13].

Here we re-examine the practical significance of the ultimate de Sitter expansion and point out a new eschatological physical consequence: from the perspective of any observer within a bound gravitational system in the far future, the static version of de Sitter space outside of that system will eventually become the appropriate physical coordinate system. Put more succinctly, in a time comparable to the age of the longest lived stars, observers will not be able to perform any observation or experiment that infers either the existence of an expanding universe dominated by a cosmological constant, or that there was a hot Big Bang. Observers will be able to infer a finite age for their island universe, but beyond that cosmology will effectively be over. The static universe, with which cosmology at the turn of the last century began, will have returned with a vengeance.

Modern cosmology is built on integrating general relativity and three observational pillars: the observed Hubble expansion, detection of the cosmic microwave background radiation, and the determination of the abundance of elements produced in the early universe. We describe next in detail how these observables will disappear for an observer in the far future, and how this will be likely to affect the theoretical conclusions one might derive about the universe.

A.1 The Disappearance of the Hubble Expansion

The most basic component of modern cosmology is the expansion of the universe, firmly established by Hubble in 1929. Currently, galaxies and galaxy clusters are gravitationally bound and have dropped out of the Hubble flow, but structures on larger length scales are observed to obey the Hubble expansion law. Now consider what happens in the far future of the universe. Both analytic [7] and numerical [10] calculations indicate that the Local Group remains gravitationally bound in the face of the accelerated Hubble expansion. All more distant structures will be

A The Return of a Static Universe and the End of Cosmology

driven outside of the de Sitter event horizon in a timescale on the order of 100 billion years ([4], see also [8, 9]). While objects will not be observed to cross the event horizon, light from them will be exponentially redshifted, so that within a time frame comparable to the longest lived main sequence stars all objects outside of our local cluster will truly become invisible [4].

Since the only remaining visible objects will in fact be gravitationally bound and decoupled from the underlying Hubble expansion, any local observer in the far future will see a single galaxy (the merger product of the Milky Way and Andromeda and other remnants of the Local Group) and will have no observational evidence of the Hubble expansion. Lacking such evidence, one may wonder whether such an observer will postulate the correct cosmological model. We would argue that in fact, such an observer will conclude the existence of a static "island universe," precisely the standard model of the universe c. 1900.

This will be true in spite of the fact that the dominant energy in this universe will not be due to matter, but due to Dark Energy, with $\rho_M/\rho_\Lambda \sim 10^{-12}$ inside the horizon volume [9]. The irony, of course, is that the denizens of this static universe will have no idea of the existence of the Dark Energy, much less of its magnitude, since they will have no probes of the length scales over which Λ dominates gravitational dynamics. It appears that Dark Energy is undetectable not only in the limit where $\rho_\Lambda \ll \rho_M$, but also when $\rho_\Lambda \gg \rho_M$.

Even if there were no direct evidence of the Hubble expansion, we might expect three other bits of evidence, two observational and one theoretical, to lead physicists in the future to ascertain the underlying nature of cosmology. However, we next describe how this is unlikely to be the case.

A.2 Vanishing CMB

The existence of a Cosmic Microwave Background was the key observation that convinced most physicists and astronomers that there was in fact a hot big bang, which essentially implies a Hubble expansion today. But even if skeptical observers in the future were inclined to undertake a search for this afterglow of the Big Bang, they would come up empty-handed. At $t \approx 100$ Gyr, the peak wavelength of the cosmic microwave background will be redshifted to roughly $\lambda \approx 1$ m, or a frequency of roughly 300 MHz. While a uniform radio background at this frequency would in principle be observable, the intensity of the CMB will also be redshifted by about 12 orders of magnitude. At much later times, the CMB becomes unobservable even in principle, as the peak wavelength is driven to a length larger than the horizon [4]. Well before then, however, the microwave background peak will redshift below the plasma frequency of the interstellar medium, and so will be screened from any observer within the galaxy. Recall that the plasma frequency is given by

$$\nu_p = \left(\frac{n_e e^2}{\pi m_e}\right)^{1/2},$$

where n_e and m_e are the electron number density and mass, respectively. Observations of dispersion in pulsar signals give [14] $n_e \approx 0.03$ cm^{-3} in the interstellar medium, which corresponds to a plasma frequency of $\nu_p \approx 1$ kHz, or a wavelength of $\lambda_p \approx 3 \times 10^7$ cm. This corresponds to an expansion factor $\sim 10^8$ relative to the present-day peak of the CMB. Assuming an exponential expansion, dominated by Dark Energy, this expansion factor will be reached when the universe is less than 50 times its present age, well below the lifetime of the longest-lived main sequence stars.

After this time, even if future residents of our island universe set out to measure a universal radiation background, they would be unable to do so. The wealth of information about early universe cosmology that can be derived from fluctuations in the CMB would be even further out of reach.

A.3 General Relativity Gives No Assistance

We may assume that theoretical physicists in the future will infer that gravitation is described by general relativity, using observations of planetary dynamics, and ground-based tests of such phenomena as gravitational time dilation. Will they then not be led to a Big Bang expansion, and a beginning in a Big Bang singularity, independent of data, as Lemaitre was? Indeed, is not a static universe incompatible with general relativity?

The answer is no. The inference that the universe must be expanding or contracting is dependent upon the cosmological hypothesis that we live in an isotropic and homogeneous universe. For future observers, this will manifestly not be the case. Outside of our local cluster, the universe will appear to be empty and static. Nothing is inconsistent with the temporary existence of a non-singular isolated self-gravitating object in such a universe, governed by general relativity. Physicists will infer that this system must ultimately collapse into a future singularity, but only as we presently conclude our galaxy must ultimately coalesce into a large black hole. Outside of this region, an empty static universe can prevail.

While physicists in the island universe will therefore conclude that their island has a finite future, the question will naturally arise as to whether it had a finite beginning. As we next describe, observers will in fact be able to determine the age of their local cluster, but not the nature of the beginning.

A.4 Polluted Elemental Abundances

The theory of Big Bang Nucleosynthesis reached a fully developed state [15] only after the discovery of the CMB (despite early abortive attempts by Gamow and his collaborators [16]). Thus, it is unlikely that the residents of the static universe would have any motivation to explore the possibility of primordial nucleosynthesis. However, even if they did, the evidence for BBN rests crucially on the fact that relic abundances of deuterium remain observable at the present day, while helium-4 has been enhanced by only a few percent since it was produced in the early universe. Extrapolating forward by 100 Gyr, we expect significantly more contamination of the helium-4 abundance, and concomitant destruction of the relic deuterium. It has been argued [17] that the ultimate extrapolation of light elemental abundances, following many generations of stellar evolution, is a mass fraction of helium given by $Y = 0.6$. The primordial helium mass fraction of $Y = 0.25$ will be a relatively small fraction of this abundance. It is unlikely that much deuterium could survive this degree of processing. Of course, the current "smoking gun" deuterium abundance is provided by Lyman-α absorption systems, back-lit by QSOs (see, e.g., [18]). Such systems will be unavailable to our observers of the future, as both the QSOs and the Lyman-α systems will have redshifted outside of the horizon.

Astute observers will be able to determine a lower limit on the age of their system, however, using standard stellar evolution analyses of their own local stars. They will be able to examine the locus of all stars and extrapolate to the oldest such stars to estimate a lower bound on the age of the galaxy. They will be able to determine an upper limit as well, by determining how long it would take for all of the observed helium to be generated by stellar nucleosynthesis. However, without any way to detect primordial elemental abundances, such as the aforementioned possibility of measuring deuterium in distant intergalactic clouds that currently absorb radiation from distant quasars and allow a determination of the deuterium abundance in these pre-stellar systems, and with the primordial helium abundance dwarfed by that produced in stars, inferring the original BBN abundances will be difficult and probably not well motivated.

Thus, while physicists of the future will be able to infer that their island universe has not been eternal, it is unlikely that they will be able to infer that the beginning involved a Big Bang.

A.5 Conclusion

The remarkable cosmic coincidence that we happen to live at the only time in the history of the universe when the magnitude of Dark Energy and Dark Matter densities are comparable has been a source of great current speculation, leading to a resurgence of interest in possible anthropic arguments limiting the value of the vacuum energy (see, e.g., [19]). But this coincidence endows our current epoch with another special feature, namely that we can actually infer the existence of both the cosmological expansion and the Dark Energy. Thus, we live in a very special time in the evolution of the universe: the time at which we can observationally verify that we live in a very special time in the evolution of the universe!

Observers when the universe was an order of magnitude younger would not have been able to discern any effects of Dark Energy on the expansion, and observers when the universe is more than an order of magnitude older will be hard pressed to know that they live in an expanding universe at all, or that the expansion is dominated by Dark Energy. By the time the longest lived main sequence stars are nearing the end of their lives, for all intents and purposes, the universe will appear static, and all evidence that now forms the basis of our current understanding of cosmology will have disappeared.

Note added in proof: After this paper was submitted we learned of a prescient 1987 paper [20], written before the discovery of Dark Energy and other cosmological observables that are central to our analysis, which nevertheless raised the general question of whether there would be epochs in the Universe when observational cosmology, as we now understand it, would not be possible.

Acknowledgment L.M.K. and R.J.S. were supported in part by the Department of Energy.

References

1. L.M. Krauss, M.S. Turner, Gen. Rel. Grav. **27**, 1137 (1995)
2. S. Perlmutter, et al., Astrophys. J. **517**, 565 (1999)
3. A.G. Reiss, et al., Astron. J. **116**, 1009 (1998)
4. L.M. Krauss, G.D. Starkman, Astrophys. J. **531**, 22 (2000)
5. A.A. Starobinsky, Grav. Cosmol. **6**, 157 (2000)
6. E.H. Gudmundsson, G. Bjornsson, Astrophys. J. **565**, 1 (2002)
7. A. Loeb, Phys. Rev. D **65**, 047301 (2002)
8. T. Chiueh, X.-G. He, Phys. Rev. D **65**, 123518 (2002)
9. M.T. Busha, F.C. Adams, R.H. Wechsler, A.E. Evrard, Astrophys. J. **596**, 713 (2003)
10. K. Nagamine, A. Loeb, New Astron. **8**, 439 (2003)
11. K. Nagamine, A. Loeb, New Astron. **9**, 573 (2004)
12. J.S. Heyl, Phys. Rev. D **72**, 107302 (2005)
13. L.M. Krauss, R.J. Scherrer, Phys. Rev. D **75**, 083524 (2007)
14. A.G.G.M. Tielens, in *The Physics and Chemistry of the Interstellar Medium* (Cambridge University Press, Cambridge, 2005)
15. R.V. Wagoner, W.A. Fowler, F. Hoyle, Astrophys. J. **148**, 3 (1967)
16. R.A. Alpher, H. Bethe, G. Gamow, Phys. Rev. **73**, 803 (1948); R.A. Alpher, J.W. Follin, R.C. Herman, Phys. Rev. **92**, 1347 (1953)
17. F.C. Adams, G. Laughlin, Rev. Mod. Phys. **69**, 337 (1997)
18. D. Kirkman, D. Tytler, N. Suzuki, J.M. O'Meara, D. Lubin, ApJ Suppl. **149**, 1 (2003)
19. S. Weinberg, Phys. Rev. Lett. **59**, 2607 (1987); J. Garriga, M. Livio, A. Vilenkin, Phys. Rev. D **61**, 023503 (2000)
20. T. Rothman, G.F.R. Ellis, Observatory **107**, 24 (1987)

Glossary

Anti-DeSitter Space Space-time with constant negative curvature.
Baryonic Matter Known matter (as opposed to Dark Matter).
Big Bang The term is used to refer to the singularity at the beginning of our Universe. The Big Bang theory explains how the Universe is expanding from its initial state.
Big Crunch Reversal of the Big Bang, where the whole Universe collapses to a singularity.
Big Rip A scenario where the acceleration of the expansion increases with time, resulting in a "tearing" apart of the very fabric of spacetime.
Black Hole A region of space-time that is bended inward due to the extreme force of gravity. It traps anything even light that passes its event horizon.
Brown Dwarfs Brown dwarfs are astronomical objects that are too small to sustain hydrogen fusion in their cores.
Chandrasekhar Mass A mass threshold in stellar structures named after the Indian astrophysicist. When the Chandrasekhar mass limit is breached, the degenerate star is too heavy to support itself, and the object blows up in a supernova.
Concordance Model of Cosmology Concordance Model of Cosmology is a homogeneous and isotropic solution of Einstein's theory of gravitation with a cosmological constant and with vanishing curvature.
Cosmic Microwave Background Remnant E&M radiation from big bang. It is now at a temperature of about 2.7 K (2.7° above absolute zero).
Cosmic Strings These are different objects from strings in string theories. Cosmic strings are theoretical objects that have been developed at the beginning phase of the Universe. Extremely long and narrow but massive, they could be stretched across the Universe.
Cosmic Topology Using topology to construct cosmological models.
Cosmology Study of the cosmos that deals with the origin, the evolution, and the fate of the Universe.

Cosmological Constant A term used by Einstein in his general relativity equation for the purpose of counter effecting gravitational attraction. Based on observed acceleration in the expansion of the Universe, it has a small value of the order of $10^{-29}\,\mathrm{g\,cm^{-3}}$.

Cosmological Heat Death A term that refers to a possible state of the Universe in which it expands adiabatically (no new entropy is produced).

Cosmological Principle The strong version states that locally the Universe is isotropic about every point and hence homogeneous. The weak version requires this only for the average distribution on large scales.

Dark Energy Dark Energy is believed to be the vacuum energy with negative pressure in its simplest form that causes the Universe to accelerate. It accounts for about 75% of the matter/energy of the Universe.

Dark Matter A kind of matter that cannot be seen directly and its composition is unknown. However, its gravitational effect can be measured. It responds to gravitational force, but it does not respond to strong, weak, and electromagnetic forces. Dark Matter accounts for 22–25% of the total energy of the Universe.

Dark Matter Halos Large structures that extend far beyond the visible portions of galaxies marked by stars, gas, and other forms of ordinary matter.

De Sitter Space In homogeneous cosmology this term is also used to refer to a flat space with no matter and a cosmological constant

DeSitter Universe If the Universe ends up in a cosmic heat death, cold and empty of structure, maintaining its entropy and temperature at constant values eternally. Such a Universe is known as a DeSitter Universe.

Doppler Effect The shift in frequency of a wave (to a higher frequency, when its source is moving toward a receiver; and to a lower frequency, when the source is moving away from the receiver) for an observer moving relative to the source of the wave.

Entropy A measure of the disorder or chaos of a closed system.

Event Horizon An event horizon is a hyper-surface in space-time beyond which events cannot affect an outside observer.

Exotic Material Hypothetical material inside of a wormhole which has negative average energy density.

Flatness Problem One of the three problems associated with the standard models of cosmology. The problem has to do with the basic question, why the Universe is close to being spatially flat.

Friedmann–Lemaitre–Robertson–Walker Models Spatially homogeneous and isotropic models of cosmology.

General Theory of Relativity It formulates how gravity bends space-time, and it is used to explain and understand the large-scale structure of the Universe.

Halos It is believed that galaxies and galaxy clusters are embedded in giant halos of Dark Matter.

Hawking Effect A slow quantum mechanical process that ultimately leads to the decay of black holes. Through quantum effects, virtual particles are created near the event horizon of a black hole. Although such particles only live for a short time, the tidal stretching force, which is enormous near a black hole, does work on them while they remain in existence. If the work done on the particle – by the tidal force – is large enough, the particle is promoted from virtual existence to "real" existence. The particle can then leave the black hole and is thus effectively emitted by the hole.

Hawking Radiation Radiation produced by the Hawking effect.

Hawking Temperature The Hawking temperature of a black hole is given by the expression:

$$T_{bh} = \frac{hC^3}{8\pi k G M_{bh}}$$

where M_{bh} is the mass of the object. Note that this temperature is extremely small: $T_{bh} \approx 6 \times 10^{-8}$ K for a black hole with one solar mass.

Homogeneous Everywhere the same.

Horizon Problem One of the three problems associated with the early models of bing bang cosmology. The problem has to do with the causal connection of different parts of the Universe.

Hubble Constant H The constant in Hubble's law which is used to calculate the size and age of the Universe.

Hubble Expansion Law Simply stated by $V = HD$ equation, where V is the recession velocity of objects such as galaxies, H is the Hubble constant, and D is distance from Earth.

Hubble Time Hubble time is comparable to the current age of the Universe.

Inflationary Cosmology It states that the Universe (the space) went through exponential expansion very early after the big bang.

Isotropic The same in every direction. In cosmology, it means the Universe looks the same in every direction.

LISA Laser Interferometer Space Antenna. http://lisa.nasa.gov/.

LHC Large Hadron Collider. It is the world's largest and highest energy particle accelerator. A circular (27 km in circumference) particle accelerator (proton–proton collider) laboratory at CERN in Geneva.

MACHOS Massive Astrophysical Compact Halo Objects.

Multiverse This term is used to refer to Multi-Universes.

Nebulae A Nebulae is an interstellar cloud of dust and gases. This term was also used in early observations and they turned out to be other galaxies beyond our Milky Way galaxy.

Neutrino An elementary particle that is electrically neutral (does not carry electric charge).

Neutron Stars The star, with roughly the mass of the Sun and a radius of only 10 km, becomes essentially one gigantic nucleus, with most of the electrons and protons combining to make neutrons. Hence the name "neutron stars."

No Hair Theorem A constraint which states that only three properties of a black hole can be observed outside its event horizon. (The black hole mass. The spin of the black hole. The electric charge of the black hole).

Planck Length A quantity associated with quantum gravity. It is about 10^{-35} cm.

Planck Mass A quantity associated with quantum gravity. It is about 10^{-8} kg.

Planck Time A quantity associated with quantum gravity. It is the time it takes for light to travel a Planck length interval which is about 10^{-42} s.

Quantum Cosmology A branch of cosmology that uses the laws of quantum mechanics to study the cosmos.

Quantum Gravity A theory that unifies General Theory of Relativity with quantum mechanics under one single framework.

Red Dwarfs Stars belonging to the smallest class of stars which live much longer compared to other stars are known as "M stars" or "red dwarfs."

Red Giant A star of low or intermediate mass will become a Red Giant in its late phase of stellar evolution. The Sun will become a red giant in about seven billion years from now.

Second Law of Thermodynamics Entropy of a closed system is always greater than or equal to 0. It cannot decrease.

Smoothness Problem One of the three problems associated with the standard models of cosmology. The problem has to do with the basic question why the matter is uniformly distributed in the Universe.

Singularity A point in space-time where its curvature becomes infinite. Big Bang is an example of a singularity.

Special Theory of Relativity It states that laws of nature are the same for all observers regardless of how they move. Also, it describes that space and time are connected and no longer individually absolute.

Standard Model of Cosmology Standard model of Cosmology consists of the following theories and models: General Theory of Relativity, Dark Matter, Dark Energy, initial conditions at Big Bang, and Standard model of particle physics.

Stellar Black Holes These objects have masses in the range of 10 to perhaps 100 Suns.

Supermassive Back Holes Astronomical observations clearly show that almost every large galaxy contains a monster black hole at its core. These black holes come in a range of masses, from about one million to one billion times the mass of our Sun.

Supernova/Supernovae A supernova is a nuclear stellar explosion at the end of the star's life.

Thermodynamic Arrow of Time The thermodynamic arrow of time is based on the Second Law of Thermodynamics that in a closed/isolated system, entropy increases with time.

Thermodynamical Equilibrium A system with maximum amount of entropy.

Topology A branch of mathematics that deals with spatial properties that are preserved under continuous deformations of objects.

Uncertainty Principle One of the basic principles of quantum mechanics developed by W. Heisenberg. It formulates that one cannot precisely specify the values of two conjugate terms such as position-momentum or time-energy.

White Dwarfs A white dwarf is a very dense small star. Approximately 997 of every 1,000 stars will turn into white dwarfs upon their death. These stellar remnants typically retain somewhat less mass than that of our Sun, but they are much smaller in radius and are one million times denser.

WIMPS Weakly Interacting Massive Particles.

White Hole The time reversal of a black hole. Big bang is an example of a white hole. A theoretical region of space-time where matter erupts but cannot enter the region.

About the Authors

Fred C. Adams is Professor of Physics at The University of Michigan, Ann Arbor. He received his PhD in Physics from the University of California, Berkeley, in 1988. For his PhD dissertation research, he received the Robert J. Trumpler Award from the Astronomical Society of the Pacific. After serving as a postdoctoral research fellow at the Harvard-Smithsonian Center for Astrophysics (Cambridge, MA), he joined the faculty in the Physics Department at the University of Michigan (Ann Arbor, MI) in 1991. Adams was promoted to Associate Professor with tenure in 1996, and to Full Professor in 2001. He is the recipient of the Helen B. Warner Prize from the American Astronomical Society and the National Science Foundation Young Investigator Award. He has also been awarded both the Excellence in Education Award and the Excellence in Research Award from the College of Literature, Arts, and Sciences at the University of Michigan. In 2002, he was given The Faculty Recognition Award from the University of Michigan. He has recently been named to as a Senior Fellow for the Michigan Society of Fellows. Professor Adams works in the general area of theoretical astrophysics with a focus on the study of star formation and cosmology. He is internationally recognized for his work on the radiative signature of the star formation process, the dynamics of circumstellar disks, and the physics of molecular clouds. His recent work in star formation includes the development of a theory for the initial mass function for forming stars and studies of extra-solar planetary systems. In cosmology, he has studied many aspects of the inflationary universe, cosmological phase transitions, magnetic monopoles, cosmic rays, anti-matter, and the nature of cosmic background radiation fields. His recent work in cosmology includes a treatise on the long-term fate and evolution of the universe and its constituent astrophysical objects.

Thomas Buchert is Professor of Cosmology at the University Claude Bernard in Lyon, France. He is a leading expert in the research field on inhomogeneous cosmological models. He worked as Research Associate at the Max-Planck-Institute for Astrophysics in Garching, Germany, in the period 1984–1995 during which he obtained his PhD in Theoretical Physics from the University of Munich in 1988. During the period 1988–1994, he took several short-term visiting positions in

Europe being Member of the European Cosmology Network. His research was focussed on cosmological structure formation theories, where the heart of this work was defined within a 5 years project of the German Science Foundation leading to his Habilitation in Astronomy, received from the University of Munich in 1994. He organized exchange projects with France and Spain, and he was active in the Max-Planck exchange programme with the Chinese Academy of Sciences. In 1995, he obtained the degree Lecturer at the University of Munich. Since then until 2006 he worked as Research Associate at the Technical University in Munich, and as a Lecturer in Theoretical Physics and Cosmology. During that time he was leading a research group within a project on Astroparticle Physics as PI on morphological statistics of cosmic structure. From 1998 until 2006 he took several long-term visiting positions as Associated Member of Personnel at CERN in Geneva, Switzerland, as Tomalla Visiting Professor at University of Geneva, as Center of Excellence Researcher at the National Astronomical Observatory in Tokyo, Japan, and as Monkasho Invited Professor at the University of Tokyo at the Research Center for the Early Universe, during which he also worked as Visiting Professor at Tohoku University in Sendai, Japan, and the Tokyo Institute of Technology. During the summer term 2006, he took a temporary Chair as Full Professor in Theoretical Physics at the University of Bielefeld, Germany. Since then he regularly worked at the Observatory of Paris in France, became Staff Member at the Observatory of Lyon, and Full Professor at the University Claude Bernard, Lyon, in 2007. He gives courses on gravitational theories, mathematical physics, kinetic theory, and cosmology within the Master Programme at École Normale Supérieure in Lyon. Since 2010, he is head of a large team dealing with galaxy physics, simulations, instrumentation projects, and theoretical cosmology, and he is PI of a collaboration on Dark Energy and Dark Matter. Professor Buchert works in the areas of theoretical, observational, and statistical cosmology with a focus on the understanding of global properties of world models. His research interests also include Riemann–Cartan geometry, integral geometry, and nontrivial topologies of spaceforms. He is internationally recognized for innovations on the morphological analysis of galaxy catalogues and Cosmic Microwave Background maps, on the foundations of the Lagrangian theory of structure formation, and for a set of equations governing the average evolution of cosmological models in general relativity and their implications for an explanation of the Dark Energy and Dark Matter problems.

Laura Mersini-Houghton is Professor of Cosmology and Theoretical Physics at UNC-Chapel Hill. She did her bachelor's degree at the University of Tirana. Then she received a Fulbright Scholarship to study at the University of Maryland-College Park, where she received her Master's degree in 1997. She then moved to the University of Wisconsin–Milwaukee where she finished her Ph.D. under the mentorship of L. Parker in 2000. She was awarded a postdoctoral research grant at Scuola Normale in Pisa during 2000–2002. She joined the faculty at the University of North-Chapel Hill in 2004 and was promoted as Associate Professor with tenure in 2008.

Her main research areas are foundational issues related to the early universe and the current acceleration of the universe. She proposed a theory of the initial conditions of the universe soon after the discovery of the landscape of string theory around 2004. Her theory assumes that before the Big Bang, the universe is a wavefunction propagating in a landscape of possible Big Bangs. As such the theory strongly advocates the existence of a multiverse. It is the only theory that shows why the universe(s) can only start at high energies. Three of the predicted signatures in the sky of this theory for the birth of the universe from the landscape have been tested recently. For this reason, her theory on how the universe started has received worldwide media attention and has been featured in many science magazines and TV programs such as BBC-Horizon, National Geographic, and "Through the wormhole: with Morgan Freeman."

Her previous work on the current acceleration of the universe explored the possibility that the fabric of spacetime at very short distances, which is the realm of quantum gravity, obeys a different relation between the energy and velocity of modes. As the universe grows, this fabric gets stretched (redshifted) and so do the short distance modes along with it. Since these modes are not short any longer, they contribute to the Dark Energy in the universe. This process of replenishing the energy of the universe by the short distance modes that enter it due to being redshifted continues ad infinitum.

Index

A
Acceleration, 22–23, 28, 29, 38–40, 42, 48, 52, 54, 64, 74, 77
Andromeda, 74, 82, 83, 116, 120
Anisotropic, 41
Architecture, 1, 13–15, 31, 33, 89
Autocorrelation function, 37
Average, 8, 13–19, 22, 27, 28, 31, 39–44, 46–48, 89

B
Background, 5, 12–17, 35–38, 67, 71, 76–81, 86, 87, 108, 114
Background space, 13, 78, 114, 116
Backreaction, 22, 23, 41–48, 55–61, 66
Baryonic matter, 39
Big Bang, 4–6, 52, 72, 75–80, 103
Big Bang nucleosynthesis, 77, 122
Big Crunch, 4, 52, 57–63
Big Rip, 52, 80–81, 116, 125
Biological system, 32
Black holes, 11, 17, 55, 80, 91, 92, 95, 97, 100–106, 114–115
Blue excess, 29
Boltzmann constant, 107
Boundary, 2, 5, 6, 14, 33, 34, 54, 63, 64, 113
Braneworld, 13
Brown Dwarfs, 84, 91, 92, 94, 97, 104, 110, 125
Bruno, G., 6

C
Carfora, M., 6, 16, 25
Catenary model, 4
ΛCDM Universe,

Center for Astrophysics (CfA) Survey, 7, 8
Chandrasekhar limit, 91, 92
Chandrasekhar mass, 94, 125
Coma cluster, 8
Commutation, 17
Concordance model, 6, 12, 19, 20, 37, 39. *See also* Standard model
Conservation law, 42
Construction principle, 26–32
Copernicus, N., 71, 72
Cosmic isolation, 81–82
Cosmic microwave background (CMB), 5, 12, 16, 35–38
  dipole, quadrupole, multipole, 36
  topology of, 35–36
Cosmic topology, 32, 125
Cosmological arrow of time, 54, 64
Cosmology
  constant, 4, 6, 12, 22, 23, 28, 31, 32, 39, 41, 43, 44, 46–48, 51–67
  equations, 18, 22, 39–45, 47–48
  parameters, 39, 40, 71
  physical, 10, 12
  strong principle, 27
  weak principle, 27
Covering space, 33, 34
Curvature
  extrinsic, 24, 44
  intrinsic, 24, 44
  Ricci, 24, 44
  Riemann, 24, 44
  Weyl, 24, 44

D
Dark Energy, 1–48
Dark Matter, 1–48, 64, 81, 86, 93–95, 104, 106, 109, 111, 112

Dark mysteries, 11, 25
Dark sectors, 9–12
Degenerate era, 91, 93, 104
Degenerate objects, 90–94, 116
De Sitter horizon, 53, 55, 57, 59
De Sitter hubble time, 51, 54, 60, 63
De Sitter model, 4
De Sitter space, 119, 120, 126
De Sitter state, 4
De Sitter Universe, 120
De Sitter, W., 3
Distance scale, 7
Doppler effect, 36
Doppler shift, 126
Dynamical relaxation, 109

E
Early Universe, 11, 76, 77, 101, 105, 123, 124
Eddington model, 4, 48
Eddington, Sir Arthur, 3–6, 28, 29, 33
Effective sources, 43, 48
Einstein, A., 3–7, 9, 17, 18, 28–31, 33, 39, 47–48
Einstein cosmos, 4, 5, 31, 34, 47, 48. *See also* Universe, static
Einstein's equations, 16, 19, 22, 38, 39, 66, 67
Einstein's theory of gravitation, 2, 12, 13, 17
Ellis, G., 16
Embedding, 24, 32
Entanglement, 51, 54–56, 59, 60, 63
Entropy, 51–53, 55–57, 62–64, 72, 79, 106, 126, 128
Entropy exclusion principle, 64
Equilibrium, 28, 32, 43, 52–56, 63, 76, 78–80, 91, 103
Euclid, 19
Euclidean action, 53, 57, 62, 108
Event horizon, 51, 101, 115, 120, 125–127
Expansion, 4, 5, 7, 8, 13, 14, 22–24, 26–29, 31, 34, 38–46, 48, 51–53, 55, 57, 59, 71–77, 80, 81, 88, 106–108

F
Far from equilibrium, 32, 79
Filament, 8, 41
Four dimensional Euclidean action, 108
Friedmann, A., 3, 12, 13, 31, 38–39, 41–43, 47
Fundamental cell, 33–36
Fundamental field, 11, 12
Fundamental observer, 36
Futamase, T., 16

G
Galactic disks, 82–86, 95, 104, 109, 116
Galaxy, 7–11, 16, 26, 29, 34, 36, 37, 72–74, 76, 81–86, 88–90, 92–94, 101, 108–110, 113, 114
Galaxy catalogue, 29, 38, 123. *See also* Galaxy map
Galaxy halo, 10, 11, 80, 84, 86, 93, 109, 111
Galaxy map, 7–9, 16. *See also* Galaxy catalogue
General relativity, 3, 13, 17–18, 24, 38, 51, 54, 64, 77, 106. *See also* Einstein's theory of gravitation
Geoid, 29, 30
Geometry
  Euclidean, 13, 14, 34, 53, 57, 62, 108
  Riemann, 2, 123
Ghost, 11, 113
Gran Sasso, 11
Gravitational coupling constant, 18
Gravitational field, 21, 30, 32, 85, 93, 102
Gravitational instabilities, 51, 57, 60, 67
Gravity, 4, 11, 13, 30, 43, 51, 53, 54, 56, 60, 62–64, 77, 80, 81, 86, 93, 95, 100, 102, 103, 105, 112, 125–128, 133
Great wall, 8
GUT, 52

H
Habitable zone, 98, 99, 116
Halos, 80, 86, 104, 109, 126
Hamiltonian, 56–59
Hawking, S.W., 100
Hawking, effect, 80, 100, 102, 126, 127
Hawking evaporation, 105
Hawking radiation, 100, 102–103, 106, 115, 127
Hawking temperature, 115, 127
Heat death, 52, 78–80, 116, 126
Heisenberg uncertainty principle, 102
Hilbert space, 102
Homogeneity scale, 46
Homogeneous, 3, 4, 7, 8, 11, 13, 15–20, 27–29, 31, 32, 38–43, 45, 48, 72, 106, 107
Horizon, 26, 27, 29, 35, 51, 53–55, 57, 59, 60, 63, 67, 73, 76, 101–103, 108, 115
HORIZON simulation, 15
Hoyle, Sir Fred, 5
Hubble, E.P., 3
Hubble constant, 8, 58, 71, 107, 127
Hubble expansion, 31, 59, 120, 121, 127
Hubble radius, 60

Index 137

Hubble time, 51, 54, 60, 63, 82, 109, 127
Hypertorus, 33, 34

I

Inflation, 43, 44, 52–56, 60, 62, 67
Inhomogeneous, 8, 13, 15–20, 48, 58, 62, 122, 123
Instability, 4, 48, 56, 58, 59, 62, 63
Isotropic, 3, 4, 7, 12, 13, 27, 31, 36, 72, 106, 107

K

Kaleidoscope, 34
Kasai, M., 16
Kinematical backreaction, 41, 45, 47
Kinetic energy density, 43
Klein–Gordon equation, 44
Kolb, R., 23

L

Labini, F.S., 27
Large Hadron Collider (LHC), 11
Large-scale structure, 7, 10
Laser Interferometer Space Antenna (LISA), 64, 127
Lemaître, G., 3–6, 12–15, 28, 31
Lemaître models, 4
LHC, *see* Large Hadron Collider
Light-sphere, 36, 37
Light travel, 7, 26, 34, 82
LISA, *see* Laser Interferometer Space Antenna
Local, 13, 24–26, 28, 32, 36, 39, 54, 59, 60, 63, 73, 74, 76, 77, 82, 106, 120–123, 128
Local group, 73, 74, 76, 77, 82, 120
Luminosity, 110–113, 115

M

Massive astrophysical compact halo objects (MACHOS), 11
Matarrese, S., 23
Milky Way, 3, 7, 72–74, 82–84, 101, 103
Minisuperspace, 11, 57, 66
Minkowski space-time, 23
Morphon field, 43–44
Multipoles, 36, 56–59, 63, 102, 116
Multiverse, 1, 29, 52–54, 57, 66

N

Nebulae, 3, 7, 15
Neutrinos, 10, 39, 104
Neutron stars, 91, 92, 97, 104, 127
Newton, I., 12, 19, 21
Newtonian cosmology, 14
No hair theorem, 101, 127
Nonlocal, 59, 60, 63

O

Observable Universe, 29, 35, 73
Observational data, 10, 12, 13, 25, 37, 77, 101

P

Paradigmatic change, 25
Penrose, Sir Roger, 52
Penzias, A., 76
Perturbation. 13, 14, 26, 29, 54, 57, 58, 66, 67, 89
Phantom, 11
Picard space, 34, 35
Planck time, 72, 127
Planet scattering, 88, 89
Plasma frequency, 121
Potential energy density, 43
Proton decay, 95–97, 100, 101, 104, 106, 112, 113, 116
Pythagoras, 19

Q

Quantum entropy, 63
Quantum gravity, 64, 95, 103, 112, 127, 133
Quantum mechanics, 52, 64, 77, 102, 127, 128
Quintessence, 11, 43, 44

R

Radiation, 5, 10, 11, 36, 37, 39, 60, 67, 71, 74, 76, 79, 80, 85, 86, 95, 100, 102–104, 106, 107, 111, 112, 114, 115
Räsänen, S., 23
Recession velocity, 7
Recoherence, 54, 57, 60, 62, 63
Red Dwarfs, 89, 90, 98, 128
Red giant, 87, 88, 92, 98, 99, 114, 128
Redshift, 3, 5, 7–9, 15, 16, 29, 46, 76, 124
Ricci curvature tensor, 24
Riemann, B., 5, 17
Riemann curvature tensor, 24

S

Scalar field, 43–44
Scale-factor, 4, 38, 40, 46, 47
SDSS. *See* Sloan Digital Sky Survey
Shear, 41
Simulations, 14, 19, 21, 25, 38, 110, 113, 123
    HORIZON, 15
    Newtonian, 14, 21
Singularity, 4, 5
Slipher, V., 3, 7, 15, 75
Sloan Digital Sky Survey, 7, 9
Space
    connectivity of, 32
    elliptical, 34
    expanding, 3, 4, 28, 29, 31
    flat, 5–6
    higher-dimensional, 13
    hyperbolic, 34, 35, 39
    "outside of," 6, 73
    spherical, 5–6, 22, 30, 33, 34, 36, 39, 47
Spacetime, 51–55, 59, 67, 119, 125, 133
Space-time tube, 24, 25
Spherical topology, 33
Spherical symmetry, 26
Standard model, 6–32, 35, 38–39, 41–43, 45, 48. *See also* Concordance model
Standard model of particle physics, 11
Steady state Universe, 75–77
Stefan–Boltzmann constant, 107
Steiner, F., 35, 37
Stellar black holes, 92, 101, 115, 128
Stick man, 7, 8
String theory, 52, 133
Strong cosmological principle, 27, 128
Structure formation model, 13–15, 18, 23, 26, 33, 123
Superhorizon, 56–60, 62, 63
Superhubble, 56, 57, 59, 60, 67
Supermassive back holes, 101, 103, 115, 128
Supernovae, 22, 23, 99, 104, 115
Supersymmetric, 120

T

Tachyonic instabilities, 67
Tensor, 24
Thermodynamical equilibrium, 32
Thermodynamic arrow of time, 51, 52, 54–57, 63, 128
Time
    local, 24, 26
    universal, 24, 25
Tomita, K., 26
Topology
    "horns," 34, 35
    matching circles, 36–38
    space form, 33, 34, 36
Torus, 33, 37
Torus topology, 33

U

Uncertainty principle, 60, 62, 97, 102, 128
Universe
    empty, 23–25
    finite, 1, 2, 5, 6, 26, 31, 36, 47, 51, 56, 79
    infinite, 1, 2, 6, 31
    quasi-Euclidean, 30–31
    quasi-spherical, 31
    static, 28, 29, 36, 48 (*see also* Einstein cosmos)

V

Vacuum energy, 51–53, 55, 58, 61, 66, 67, 79–81, 107, 108, 119, 123, 126
Virgo cluster, 7
Virial equilibrium, 43
Void, 7, 8, 24, 26, 44

W

Wavefunction, 54, 55, 57–61, 64–67, 133
Wavepacket, 54, 55, 60–62, 64, 65, 67
Weak cosmological principle, 27, 30, 128
Weakly interacting massive particles (WIMPS), 11
Weyl curvature tensor, 24
White Dwarfs, 91–95, 97, 98, 104, 106, 110, 111, 113, 114, 116, 129
Wilkinson microwave anisotropy probe (WMAP), 37
Wilson, R., 76
Wiltshire, D., 26
WKB approximation, 57, 60, 62, 66

Printed by Printforce, the Netherlands